ASE Study Guide **by Chek-Chart**

S0-ADR-370

Engine Repair

FOR ASE TEST A1

Contents

Engine Repair, ASE Study Guide **by Chek-Chart**
Copyright © 1998 by Chek-Chart Publications

All rights reserved. Printed in the United States of America. No part of this publication may be reproduced, stored in a retrieval system, or duplicated in any manner without prior written permission of the publisher.

International Standard Book Number: 1-57932-092-9

Printed in the United States of America.

Chek-Chart Publications

320 Soquel Way
Sunnyvale, CA 94086
408-739-2435
"Chek" us out on the web at www.chekchart.com.

Editor
William J. Turney, CMAT, L1

Contributing Editor
Richard K. DuPuy

Technical Consultant
Jerry Mullen

Production Team Supervisor
Tricia Flodder

Production Team
Svetlana Dominguez
Maribeth Echard
Kristy Nash
Carl Pierce
Mary Ellen Stephenson

DISCLAIMER

Chapter One

GENERAL ENGINE DIAGNOSIS

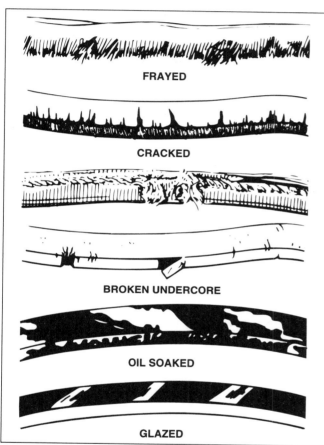

1-1 *Inspect drive belts for wear and damage.*

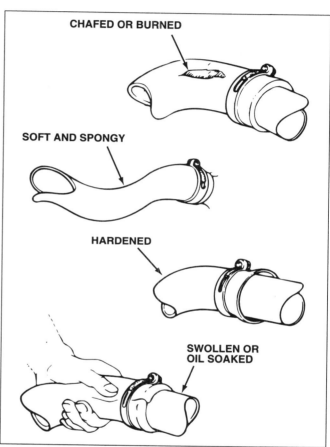

1-2 *Check hoses for damage.*

ENGINE INSPECTION AND DIAGNOSIS

Before you begin any engine repair, it is necessary to identify, as closely as possible, the defects involved. This chapter explains the tests that are performed to identify problem areas, while the engine is still in the vehicle and in operating condition.

The most important aspect of engine diagnosis is listening to the customer. Let the customer describe the symptoms and clue you in to the problem. This will give you a fairly good idea of where to begin your search. Then, conduct a visual inspection of the vehicle and verify the complaint by experiencing the symptoms during a test drive.

Preliminary Inspection

A quick inspection before your test drive can often isolate the source of a problem. Before you drive the vehicle and begin engine testing, do the following:

- Check all drive belts to make sure none are cracked, frayed, loose, or slipping, figure 1-1.
- Check electrical harnesses and connectors for loose contacts, brittle insulation, and broken wires.
- Check all engine-mounted components for loose or missing bolts, worn bushings, and loose or broken support brackets.
- Check all hoses to make sure none are soft, brittle, kinked, or otherwise damaged, figure 1-2.

- Check all fuel lines, hoses, and fittings for leakage, loose connections, and damage.
- Check the battery; look for corrosion on the terminals, proper electrolyte level, and missing or damaged holddown assemblies.
- Check the secondary ignition cables for damaged insulation, corrosion, and loose connections.
- Check the air cleaner filter element for dirt or blockage.
- Check emission control systems for brittle, burned, or damaged hoses and loose fittings.

If no obvious problems are found, continue your evaluation with a test drive.

Test Drive

Whenever possible, let your customer accompany you on the test drive. The customer knows the vehicle and can point out sounds, vibrations, and other annoyances that you might assume to be normal or overlook.

Conduct a thorough road test; one quick lap around the block is not enough. The engine must be brought to normal operating temperature, and your test should include a stop-and-go city driving cycle and a period of cruising at highway speed. If the car has air conditioning, operate it during the test drive. If a hilly stretch of road is available, include it in your route.

Unusual Engine Noises

Engine noises can be divided into two general categories: those that originate in the top end of the engine, and those that originate in the bottom end of the engine. Begin engine noise diagnosis by determining where in the engine the noise is coming from. Bottom end, or crankcase, noises occur at crankshaft speed, so they tend to produce a high-frequency knock or rumble. Top end, or valve train, noises occur at a lower frequency because these parts operate at one-half crankshaft speed.

A stethoscope is a handy tool for isolating noises. You can also use a timing light to determine whether a noise is from the top or bottom end of the engine. Connect the timing light and listen. If the engine noise cycles in time with the flashing light, the sound is coming from the bottom end. Sounds that are audible with every other flash of the timing light originate in the top end of the engine.

Top-end noises

The top end of a healthy engine produces a high-pitched, whirring noise with a very rapid and much fainter sewing-machine-like clicking coming from the valves. The more valves the engine has, and the higher the idle speed, the more the individual clicks will blend into a constant drone. Any deviation is abnormal and indicates a problem. Listen for:

- An irregular clacking or knocking noise caused by excessive camshaft endplay.
- An irregular slapping or thumping at the front of the engine caused by a loose timing belt. A tight timing belt makes a whirring, whining, hum that rises and falls in pitch with rpm.
- A single, clear clack whenever a particular valve opens can be a collapsed lifter or a broken valve spring.
- A loud, cycling, valve rattle that you can hear over the normal valve noise can indicate either worn valve guides or rocker arm pivots.
- Low pressure or restricted oil flow will produce an excessively loud, rhythmic clatter.

Bottom-end noises

Healthy engines produce an evenly pitched, rapid, whirring sound and nothing else. Knocking or thumping noises are signs that something is wrong. In general, bottom-end noise can be caused and indicated by:

- An irregular knock at idle that can be made louder or fainter by playing with the clutch pedal indicates too much crankshaft endplay.
- A sharp clattering knock that may be continuous at idle or only appear when the throttle closes suddenly, can indicate a bad connecting rod bearing. The noise will diminish if the spark plug for the offending cylinder is grounded.
- A hollow metallic clatter that is loudest when the engine is cold may be piston slap caused by too much piston to cylinder wall clearance. Grounding the spark plug of the affected cylinder will often make piston slap louder because it eliminates the cushioning of the extra gas pressure pushing on the piston.
- A sharp knocking that stands out most at idle can indicate a **wrist pin** that is loose in its bore. Grounding the spark plug of the affected cylinder makes the knock audible at top dead center as well as bottom dead center. Retarding the spark decreases wrist pin noise.
- A rapid, steady dull pounding that increases with load is typical of worn main bearings.

Unusual Exhaust Color and Odor

Although a healthy catalytic converter can do a good job of cleaning up the exhaust, you can tell something

---

Wrist Pin: The cylindrical or tubular metal pin that attaches the piston to the connecting rod (also called the piston pin).

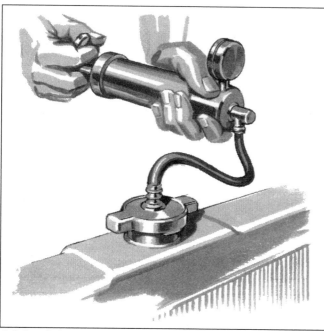
1-3 *Attach the pressure tester to the filler neck.*

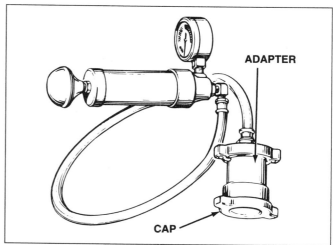
1-4 *Radiator cap pressure test.*

about the internal engine condition by checking for unusual smoke or smells:

- Black exhaust smoke. This is caused by a rich air-fuel mixture, and is often accompanied by the "rotten-egg" smell of an overworked catalytic converter.
- Blue exhaust smoke indicates excessive oil burning, which gives off a pungent odor.
- Cloudy white exhaust is often the result of engine coolant leaking into the combustion chamber. Burning coolant also produces a distinctive chemical odor. Check the temperature gauge for overheating.

Keep in mind that oil vapor odors are not always the result of an internal engine problem. A clogged or malfunctioning positive crankcase ventilation (PCV) system can not only produce a burning-oil smell, but can also cause excessive crankcase vapor and increase oil consumption. Always check all external sources before you condemn the engine.

Inspecting for Fluid Leaks

Fluid leaks, especially engine oil, are often difficult to locate because before the problem is detected, the leaking fluid has spread over a large area. Therefore, it is often advantageous to steam clean or pressure wash the engine before you try to isolate the source of a leak.

Coolant leaks

The quickest way to find an external coolant leak is by pressure testing the cooling system. Perform the test on a cold engine using a hand pump with a pressure gauge as follows:

1. Remove the pressure cap and attach the radiator pressure tester to the filler neck.
2. Pump the tester until the gauge reading matches the specified system pressure, figure 1-3.
3. Observe the gauge; the reading should remain steady.
4. If the gauge shows a pressure loss, pump the tester to maintain pressure and check for leaks. External leaks should be obvious as pressure forces the coolant out.
5. If there is no sign of leakage, there may be an internal leak and additional testing is required.

Your next step is to pressure test the radiator pressure cap by using the system pressure tester and an adapter, figure 1-4. Proceed as follows:

1. Attach the cap to one end of the adapter, then connect the opposite end to the tester.
2. Pump the tester until the gauge reading matches the pressure rating of the cap.
3. Observe the gauge; it should hold steady within 1 or 2 pounds per square inch (psi) of the rating for at least 10 seconds, then gradually decrease. If the gauge reading drops too fast or too far, the cap is bad; replace it.
4. If the reading does not drop at all, continue applying pressure until the cap vents. The cap should vent when pressure exceeds the rating of the cap by 1 to 3 pounds.

Internal coolant leaks, generally the result of head gasket failure or casting cracks, are more difficult to detect. To check for leakage into the crankcase, remove the dipstick. If the oil on the dipstick looks milky or thickened, it may be contaminated by coolant.

There are several methods to check for coolant in the combustion chambers. Remove the radiator cap, start the engine, and bring it to operating temperature. A major leak will reveal itself by creating bubbles in the coolant. These bubbles rise to the surface and are visible at the filler neck. Small leaks can be detected using a chemical test kit or an exhaust gas analyzer. With chemical testing, vapors that collect above the coolant in the radiator are passed through a test fluid. The fluid is sensitive to exhaust gases and will change color if there are exhaust gasses present in the radiator.

To use an exhaust gas analyzer, hold the wand over the filler neck. Never allow coolant to be drawn into the machine. Any exhaust gasses in the system will register on the meters.

Fuel leaks

Fuel leaks are generally easy to locate simply because the system is under pressure whenever the engine is running. In addition, fuel injection systems continue to hold pressure after the engine is shut down.

Visually inspect the system. Check all fuel lines, or the fuel rail and fuel injectors, for signs of gasoline leaks. Also, check pump and fuel tank fittings.

Oil leaks

If an oil leak is severe, the source is usually obvious. Severe oil leaks generally are under pressure and come from the following locations:

- Oil sending unit
- Oil filter and gasket
- Camshaft and plug
- Crankshaft seals
- External oil lines.

Engine oil leaks that are not under pressure are more difficult to trace. The oil pan, timing cover, and valve cover gaskets are prime suspects. Check for signs of seepage along the sealing flanges. Leaks may be caused by normal wear or over tightening, but they can also be caused by too much pressure in the crankcase forcing oil past seals and gaskets. For this reason, thoroughly check the PCV system for proper operation.

Oil leaks that are difficult to find can be pinpointed by dye testing. A fluorescent dye is added to the oil and allowed to circulate in a running engine. After running, inspect the engine using a black light. The treated oil will glow brightly when exposed to the black light.

Excessive Oil Consumption

All engines burn oil during normal operation, but how much oil consumption is too much? This is always a judgment call. In general, if the engine consumes more than about 1 quart of oil every 500 miles, it needs attention. At that level, the oil burning in the combustion chamber forms significant deposits on the intake valve, piston crown, cylinder head, and spark plug.

If an engine burns oil, the spark plugs will have whitish-gray ash deposits on the electrodes. If the engine is burning a lot of oil, the plugs will be covered in liquid oil when you remove them. Oil normally enters the combustion chamber by either seeping past worn piston rings or by running down loose valve guides. An engine pulling oil through the valve guides will often have deposits built up on the intake port side of the spark plug only. Oil passing the piston rings will generally be spread more evenly on the end of the plug.

ENGINE TESTING

Specific internal mechanical problems on a running engine can be located by performing several basic tests. The tests that you will perform on a running engine include:

- Manifold vacuum
- Cylinder power balance
- Cylinder compression
- Cylinder leakage.

Manifold Vacuum Tests

As a piston moves downward in the cylinder, it draws air with it to create a pressure drop in the intake manifold. A vacuum gauge connected to the intake manifold is used to measure the difference between **atmospheric pressure** and **manifold pressure**. Vacuum gauges are usually calibrated in inches of mercury (in-Hg). Vacuum gauge readings can pinpoint:

- Intake manifold gasket and vacuum line leaks
- Valve and valve guide problems
- Retarded ignition and valve timing
- Exhaust system restrictions
- Poor combustion chamber sealing.

All vacuum readings decrease with increases in elevation, figure 1-5. Manufacturers provide specifications for testing at sea level, and it is up to you to correct these specifications for the altitude at which you are

---

Atmospheric Pressure: Weight of air at sea level, about 14.7 pounds per square inch (101 kPa), decreasing at higher altitudes.

Manifold Pressure: Vacuum, or low air pressure, in the intake manifold of a running engine, caused by the descending pistons creating empty space in the cylinders faster than the entering air can fill it.

ALTITUDE	INCHES OF VACUUM
Sea level to 1000 ft.	18 to 22
1000 ft. to 2000 ft.	17 to 21
2000 ft. to 3000 ft.	16 to 20
3000 ft. to 4000 ft.	15 to 19
4000 ft. to 5000 ft.	14 to 18
5000 ft. to 6000 ft.	13 to 17

1-5 *Vacuum gauge readings are corrected for altitude.*

testing. As a rule of thumb, subtract 1 inch for every 1000 feet above sea level.

Normal engine vacuum at idle is 15 to 21 in-Hg, figure 1-6. Vacuum decreases as the throttle opens. At steady part-throttle cruising, engine vacuum should run between 10 to 15 in-Hg. During full throttle acceleration, manifold vacuum will drop to near zero.

Vacuum tests measure intake manifold vacuum, so the gauge must be connected downstream of the throttle plates. Look for a port on the intake manifold to make the gauge connection, or tap into a vacuum line and install a tee-fitting. Avoid tapping into the canister or exhaust gas recirculation valve lines. These usually carry vacuum signals which are lower than manifold vacuum. Also, avoid lines that contain a vacuum control valve or restrictor.

Interpreting test results

Most tests are conducted with the engine idling. Interpret manifold vacuum gauge readings as follows:

- A steady reading 3 to 9 inches below normal indicates internal leakage around the piston rings, late ignition timing, or intake manifold leakage, figure 1-7.
- A reading with the gauge needle fluctuating 3 to 9 inches below normal indicates a vacuum leak in the intake system, figure 1-8.
- A leaking head gasket will cause the needle to vibrate as it floats through a range from slightly below to slightly above normal, figure 1-9.

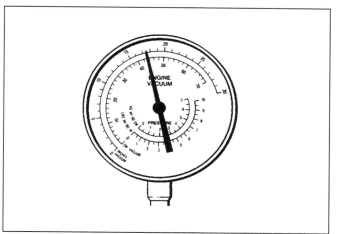

1-6 *Engine in good condition. Needle steady.*

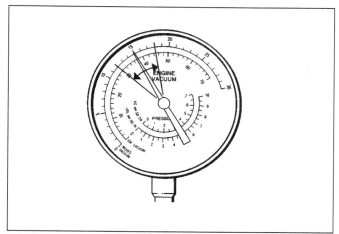

1-8 *Vacuum leak in the intake system. Needle fluctuating, low reading.*

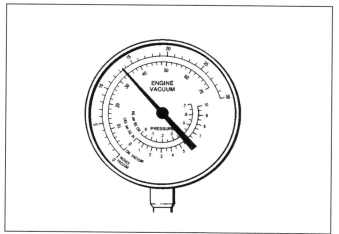

1-7 *Internal leakage or late timing. Needle steady, low reading.*

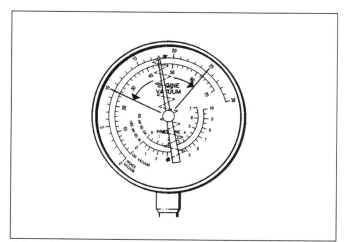

1-9 *Leaking head gasket. Vibrating needle floating low to high.*

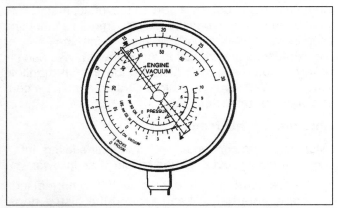

1-10 *Incorrect fuel mixture. Oscillating needle, low reading.*

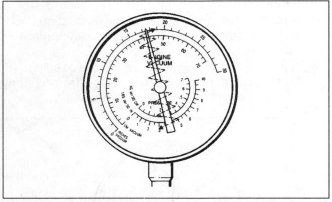

1-11 *Worn valve guides. Vibrating needle at idle, steadies at speed.*

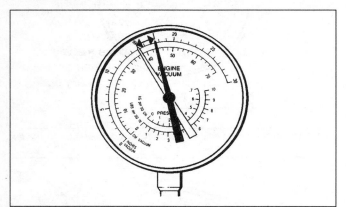

1-12 *Burned valve or valve not seating. Needle drops regularly below normal.*

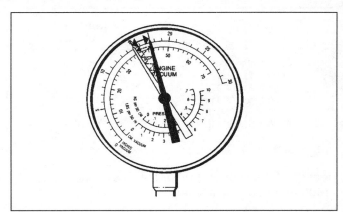

1-13 *Sticking valve. Needle drops irregularly below normal.*

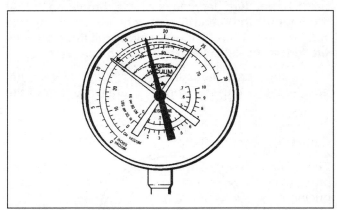

1-14 *Weak valve springs. Needle fluctuates when brought off idle.*

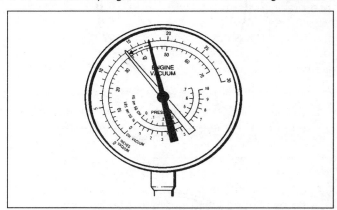

1-15 *Retarded ignition timing. Needle drops when brought off idle.*

- An **oscillating** needle 1 or 2 inches below normal could indicate an incorrect fuel mixture, figure 1-10.
- A rapidly vibrating needle at idle that steadies as engine speed increases indicates worn valve guides, figure 1-11.
- If the needle regularly drops 1 to 2 inches from the normal reading, one of the engine valves is burned or not seating properly, figure 1-12.

- An irregular drop of 1 or 2 inches from a normal reading indicates sticking valves, figure 1-13.
- Weak valve springs will produce a normal reading at idle, but the needle will fluctuate rapidly between 12 and 24 inches as engine speed increases, figure 1-14.
- A steady reading that drops 2 to 3 inches when the engine is brought off idle indicates retarded ignition timing, figure 1-15.

Oscillating: A swinging, steady, up-and-down or back-and-forth motion.

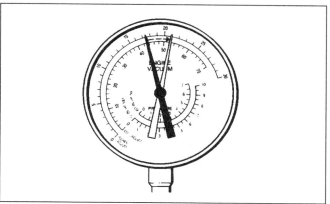

1-16 *Advanced ignition timing. Needle rises when brought off idle.*

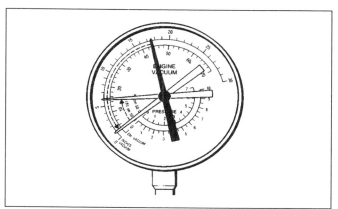

1-17 *Exhaust restriction. Needle drops at high rpm, then rises slightly.*

- A steady reading that rises 2 to 3 inches when the engine is brought off idle indicates too much ignition timing advance, figure 1-16.
- A needle that drops to near zero at high rpm, then rises slightly to stabilize at a reading below that of a normal idle indicates an exhaust restriction, figure 1-17.

Cylinder Power Balance Test

The cylinder power balance test lets you know if an individual cylinder, or a group of cylinders, is not producing its share of power. During the test, you short out the suspected spark plug, or plugs, so that there are no power strokes from the cylinder being tested. You measure results in terms of engine rpm drop, manifold vacuum drop, or a combination of these two factors. Most engine analyzers have the capability of performing a power balance test.

Perform a cylinder power balance test as follows:

1. Check the service manual of the vehicle being tested for any specific requirements.
2. Connect the engine analyzer.
3. If the engine has an EGR system, disconnect and plug the EGR valve vacuum line.
4. If the engine has electronic controls, the test should be performed while in **open-loop** operation.
5. Start and run the engine to bring it up to normal operating temperature.
6. Hold engine speed at a fast idle, about 1,000 to 1,500 rpm, or at another test speed as specified.
7. Press the button to kill one cylinder. Note and record the engine rpm drop and manifold vacuum drop. Release the button.
8. Repeat the test for each cylinder.

Interpreting test results

Compare test results for all cylinders. If the changes in engine rpm and manifold vacuum are about the same for each cylinder, the engine is in good mechanical condition. If the changes for one or more cylinders exceed 10 percent, the engine has a problem. The fault may be mechanical, or it may be in the ignition or fuel systems.

Cylinder Compression Test

A compression test reveals how well each cylinder is sealed by the piston rings, valves, cylinder head gasket, and spark plug, figure 1-18. **Compression pressure** is measured in pounds per square inch (psi) or kilopascal (kPa) using a compression gauge. The gauge measures the amount of air pressure that a cylinder is capable of producing. Vehicle manufacturers provide compression specifications for their engines in the service manual. Specifications are usually stated in

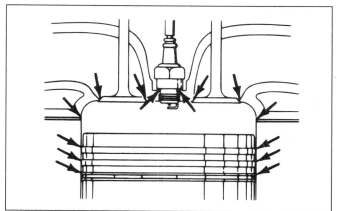

1-18 *A compression test checks how well a cylinder is sealed.*

Open-loop: An operational mode in which the engine management computer adjusts the system to function according to predetermined instructions and does not respond to feedback signals from system input sensors.

Compression Pressure: The total amount of air pressure developed by a piston moving to TDC with both valves closed.

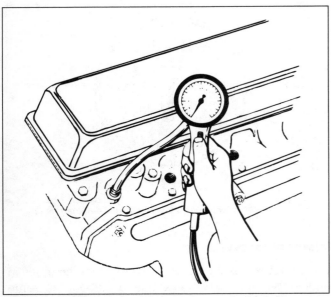

1-19 *Connect the compression gauge to a spark plug hole.*

one of two ways. Some require the lowest-reading cylinder to be within a certain percentage of the highest cylinder. Others are a minimum value, with an allowable variation between the cylinders.

Performing a compression test

For optimum results, the engine should be at normal operating temperature during testing. Remove all spark plugs prior to testing and block the linkage to hold the throttle in wide-open position. Make sure the battery is fully charged. Check compression as follows:

1. Connect the compression gauge to the No. 1 cylinder spark plug hole, figure 1-19.
2. Crank the engine for four complete compression strokes on the cylinder being tested. Watch the gauge and record readings for the first and fourth strokes.
3. Disconnect the gauge and check compression for all other cylinders.

Interpreting test results

Compare the reading for each cylinder with those specified by the manufacturer. Interpret as follows:

- Compression is normal when the gauge shows a steady rise to the specified value with each compression stroke.
- If the compression is low on the first stroke and builds up with each stroke, but not to specifications, the piston rings are probably worn.
- A low compression reading on the first stroke that builds up only slightly on the following strokes indicates sticking or burned valves.

- Two adjacent cylinders with equally low compression readings indicate head gasket leakage between the two cylinders.
- A higher than normal reading usually means excessive carbon deposits in the combustion chamber.

When compression is low, a wet test can be performed to isolate valve and piston ring problems. Squirt about 1 tablespoon of engine oil through the spark plug opening of each low-reading cylinder and retest compression. The oil acts as a temporary seal between the cylinder wall and the rings but has no effect on leaking valves or a blown head gasket. If compression increases on the wet retest, the piston rings or cylinder walls are worn. If compression does not increase on a wet test, the low compression is the result of a valve or head gasket problem.

Cylinder Leakage Test

A cylinder leakage tester, or leak-down tester, gives even more detailed results than a compression test. A leakage test can reveal:

- The exact location of a compression leak
- How serious the leak is in terms of a percentage of total cylinder compression.

Leakage testers force air into the combustion chamber through the spark plug hole. A gauge installed in the air line indicates how much pressure leaks out of the combustion chamber, figure 1-20. The gauge scale is marked from 0 to 100 percent. Cylinder leakage testers differ slightly by equipment manufacturer; you must calibrate the gauge before each test and follow the procedures for the particular unit being used.

Performing a cylinder leakage test

The following is a general procedure that will work for most equipment:

1. Start and run the engine to bring it up to normal operating temperature, then shut the engine off.
2. Remove all spark plugs, disable the ignition, block the throttle in a wide open position, disconnect the PCV hose from the crankcase, and remove the cooling system pressure cap.
3. Calibrate the tester according to instructions provided by the manufacturer.
4. Rotate the crankshaft by hand until the cylinder to be tested is at top dead center (TDC) on the compression stroke; both valves must be closed.
5. Install the tester adapter in the spark plug hole, then connect the tester to the adapter.
6. Connect the air hose to the tester and pressurize the cylinder.
7. Note the percentage of air leakage on the gauge.

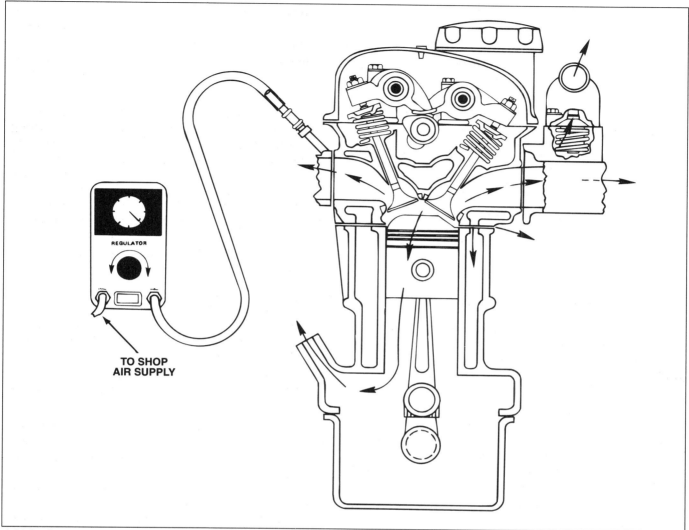

1-20 *Leakage testers use air pressure to locate compression leaks.*

Use the following table to interpret cylinder leakage tester gauge readings:

0-10%	Good
10-20%	Fair
20-30%	Poor
30-100%	Dead!

Interpreting test results

If the cylinder has more than 20-percent leakage, pinpoint the cause of the leaks as follows:

- Listen for air escaping through the air intake; this indicates a leaking intake valve.

- Listen for air escaping through the exhaust pipe; this indicates a leaking exhaust valve.
- Listen for air escaping through the crankcase and PCV system; this indicates worn or damaged piston rings, worn cylinder walls, or a worn or cracked piston.
- Watch for air bubbles in the coolant; this indicates a leaking head gasket or a crack in the engine block or cylinder head casting.
- If two adjacent cylinders have high leakage readings, the head gasket is leaking between them, or the head or block is cracked.

1. Checking manifold vacuum with a gauge will reveal all of the following problems *EXCEPT*:
   a. Worn valve guides
   b. Exhaust restriction
   c. Worn piston rings
   d. Sticking hydraulic lifter

2. A fluorescent dye test is used to check for:
   a. External coolant leaks
   b. Internal oil leaks
   c. Combustion leaking into the coolant
   d. Coolant leaking into the crankcase

3. During a manifold vacuum test a rapid needle vibration on acceleration is noticed. This can be caused by:
   a. A restricted exhaust system
   b. A rich air-fuel ratio
   c. Weak valve springs
   d. Worn valve guides

4. Low and steady vacuum gauge readings can be caused by:
   a. Worn valve guides
   b. Piston ring leakage
   c. An incorrect fuel mixture
   d. An exhaust restriction

5. An engine produces a sharp clattering knocking sound, but grounding a spark plug lead quiets the noise down. The most likely cause would be:
   a. Worn main bearings
   b. A loose wrist pin
   c. A worn connecting rod bearing
   d. A broken valve spring

6. In a cylinder compression test, if the compression is low on the first stroke and increases on following strokes but never reaches specifications, it probably means:
   a. Cylinder compression is okay
   b. Rings may be worn or valves may be sticking
   c. Head gasket may be leaking
   d. Excessive carbon build-up

7. If the manufacturer specifies 15 to 21 in-Hg of manifold vacuum at sea level, what would be the acceptable range when testing at a 5000 foot elevation?
   a. 15 to 21
   b. 20 to 26
   c. 14 to 20
   d. 10 to 16

8. Whitish-gray ash deposits on the spark plug electrodes are caused by:
   a. Burning oil
   b. Rich fuel mixture
   c. Coolant in the cylinder
   d. Retarded ignition timing

9. A cylinder compression test will detect all of the following problems *EXCEPT*:
   a. Valve not seating
   b. Worn valve guide
   c. Broken piston rings
   d. Leaking head gasket

10. High compression readings, good cylinder leakage test results, and good power balance can be the result of:
    a. Incorrect valve timing
    b. Exhaust restriction
    c. Carbon deposits on the pistons
    d. Leaking valve guide seals

Chapter Two

CYLINDER HEAD AND VALVE TRAIN DIAGNOSIS AND REPAIR

This chapter covers inspection, measurement, and repair of the cylinder head and the **valve train**.

CYLINDER HEAD GASKET REPLACEMENT

When cylinder head gasket leakage is indicated, keep in mind that head gasket failure is generally the result of another problem. Check both the cylinder head and the block for warpage, cracking, and other conditions that may have led to gasket failure.

Cylinder Head Removal

The engine must be cold to remove the cylinder head. Loosening head bolts on a warm engine can warp the head casting. The following cylinder head removal guidelines apply to most engines:

- Always disconnect the battery to avoid accidental shorting.
- For overhead camshaft (OHC) engines, the timing chain or belt must be removed from the camshaft. Secure the tensioner to prevent it from overextending.
- Air-conditioning compressors and power steering pumps can generally be unbolted from their brackets and moved aside without disconnecting lines and hoses.
- Relieve residual pressure before disconnecting fuel lines and plug the lines to prevent contamination.
- For overhead valve OHV engines, leave the rocker arms attached. Loosen them just far enough to free the pushrods. Be sure to keep pushrods in order.
- Break the cylinder head bolts loose in the reverse order of the tightening sequence. Loosen all the bolts, then remove them.
- If the head will not lift off the engine block, use a pry bar inserted into one of the ports. Do not pry on machined surfaces.

Cleaning and Inspection

Head gasket surfaces must be clean, free of scratches or surface damage, and flat within specified limits. Check the head for leakage and surface warp.

Checking for leakage

Before cleaning the sealing surface, look it over for signs of combustion, coolant, and oil leakage.

- Combustion leaks are usually obvious, because they burn away pieces of the gasket and leave a trail of carbon on the head and block surfaces.
- Coolant chemically reacts when it comes in contact with combustion. Look for traces of coolant flow etched into the metal.
- Oil leakage does not always leave a visible mark. Carefully inspect the gasket for signs of seepage, and check the oil galleries for cracking and distortion.

Cleaning surfaces

Both sealing surfaces, cylinder head and block, must be clean before a new gasket is installed.

Use a gasket scraper to remove traces of the head gasket from cast-iron surfaces. If the gasket has left an impression etched into the metal, clean it off with a block of wood or a file wrapped with emery cloth. A scraper can easily gouge aluminum surfaces. When cleaning aluminum cylinder heads, use only the block and emery cloth. You can also clean an aluminum gasket surface safely with nylon-mesh pads that chuck into a drill motor.

Checking for warpage

To check a surface for flatness, place a straightedge across it, figure 2-1. Try to slip a feeler gauge under it to determine if the head is warped. Check the casting horizontally, vertically, and by opposing corners in an

---

Valve Train: The assembly of parts that transmits force from the cam lobe to the valve.

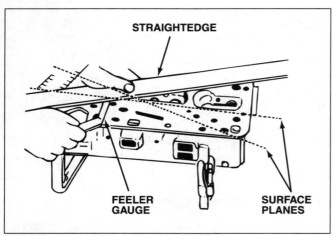

2-1 *Check surface flatness with a straightedge.*

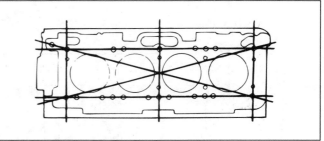

2-2 *Reposition the straightedge several times to get an accurate picture of surface condition.*

"X" pattern, figure 2-2. Warpage is equal to the largest feeler gauge that will slip under the straightedge without lifting it.

Cylinder Head and Gasket Installation

If the cylinder head is in sound condition, it is ready to be installed. If problems are found, recondition or replace the cylinder head.

Cleaning threaded holes and bolts

Use a bench grinder with a wire wheel to clean head bolts. Thoroughly clean the threaded portion and the seating area on the underside of the bolt head.

Run a thread chaser or a bottoming tap down the threads of each bolt hole. Remove the tool and clear any residual debris out of the hole with a blast of compressed air.

Installing the head gasket

Many head gaskets are marked "top" or "front" to indicate which way they install. Make sure any directional markings are properly oriented.

Check fit the gasket. Be sure all locating dowels are in position. Then, fit the gasket over the dowels and onto the block. Head gaskets install without sealer unless otherwise specified by the engine manufacturer.

Installing the cylinder head

Slowly lower the cylinder head straight down onto the block to engage the dowels and lock it in place. Install the head bolts and run them in finger tight, then tighten incrementally and in sequence according to specifications.

Torque-to-yield bolts, also known as stretch bolts, may be used to attach the cylinder head. Most torque-to-yield bolts are not reusable and must be replaced whenever they are removed. Torque-to-yield bolts generally have a reduced diameter shank, unusual head configuration, or other distinctive characteristics to make them recognizable, figure 2-3.

Torque-to-yield bolts are brought to torque, then tightened an additional amount by a specific number of degrees. A torque angle gauge is used to measure degrees during final tightening, figure 2-4.

CYLINDER HEAD DISASSEMBLY, CLEANING, AND INSPECTION

Before removing the valves, measure the original valve spring **installed height**, so you can restore proper spring tension and correct rocker arm angle on reassembly. Installed height is the distance from the spring seat, or cylinder head casting, to the bottom of the spring retainer, figure 2-5.

Removing Valves, Springs, and Keepers

An accumulation of varnish and sludge forms between the valve stem, keepers, and spring retainer as a normal result of engine operation. This causes the parts to stick together and can make them difficult to remove. Lightly tapping the top of the valve spring with a hammer will generally loosen the parts.

Use a valve spring compressor to depress the valve spring, then remove the keepers. A small magnet works well for lifting the keepers off of the valve stem. Once the keepers are removed, release the valve spring compressor and lift off the valve spring, retainer, and shims. Then, remove the valve and valve stem seal. Set the parts aside, but keep them in order for inspection and reassembly.

Installed Height: The dimension from the valve spring seat to the bottom of the spring retainer, also called assembled height.

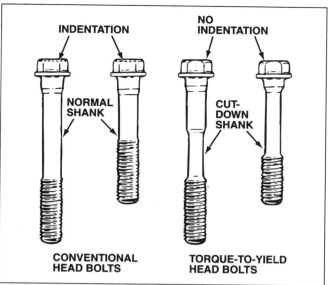

2-3 *Torque-to-yield bolts have distinguishing characteristics for easy identification.*

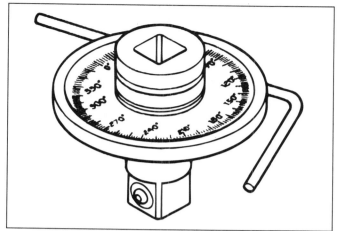

2-4 *A torque angle gauge gives accurate degree readings for final tightening.*

Cylinder Head Cleaning

Most cleaning methods require that all cover plates, sending units, core plugs, and other attachments be removed from the cylinder head. In addition, some or all of the manifold studs may have to come off if resurfacing is required. Strip the head to a bare casting to prepare for cleaning.

Cleaning methods

Cylinder heads can be cleaned using a variety of methods. Common cleaning equipment includes:

- Hot tanks
- Spray booths
- Glass bead blasters
- Airless shot blasters
- Thermal ovens.

After cleaning, some carbon will probably still cling to the cylinder head and back sides of the valves. Use a rotary wire brush that chucks into a drill motor to finish cleaning the ports and combustion chambers.

Scale deposits that build up in the coolant passages can be broken loose with a thin, flexible scraper and cleaned out with a small wire brush. Flush the passages with solvent, then clear and dry with compressed air.

Run a rifle brush completely through the oil galleries to remove any built-up deposits and trapped debris.

Evaluating Head Condition

Once the cylinder head casting is clean, inspect it for warpage, casting cracks, and other signs of damage.

Checking for warpage

The procedure for checking the head gasket surface was detailed previously. Recheck now that the head is clean. Surface warpage, if not too severe, can be corrected by machining. Flatness and finish can be restored either by **surface grinding** or by **milling**.

Warped aluminum heads can be straightened by heat treating. The head is bolted to a surface plate with shim stock strategically placed between the head and

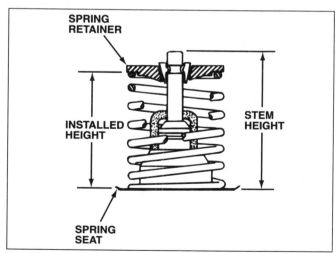

2-5 *Measure valve installed height before disassembling the cylinder head.*

Surface Grinding: The process of using a power-driven abrasive stone to remove metal from a casting to restore the surfaces.

Milling: The process of using a multiple-tooth cutting bit to remove metal from a workpiece.

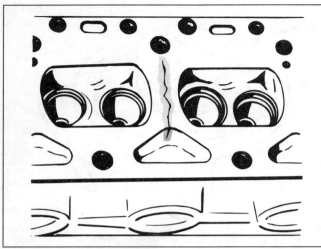

2-6 *Magnetic particles accumulate along a crack to make it more visible.*

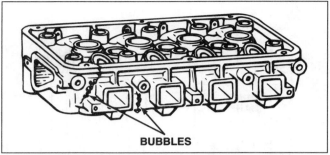

BUBBLES

2-7 *Pressure testing will cause the soap solution to bubble when air escapes.*

plate. The assembly is then heated and cooled in an oven to relieve tension in the casting.

The alignment of the camshaft-bearing journals should be checked on an overhead cam cylinder head. Lay a straightedge across all of the camshaft-bearing journals. It should rest firmly in place and contact each of the journals. If not, use a feeler gauge to measure the warpage. Most heads will bow up in the center when they warp. Actual warpage will equal one half the thickness of the largest feeler gauge that will fit between the straightedge and bearing saddle.

Inspecting for cracks

Several methods are available for locating casting cracks. These include:

- Magnetic particle inspection
- Pressure testing
- Dye penetrant testing.

Magnetic particle inspection

Magnetic particle inspection can be used only on ferrous metals (cast iron), and will reveal cracks only on visible surfaces of the casting. There are two types of magnetic particle test medium—dry powder and fluorescent liquid. Both mediums are magnetic. However, the powder is visible to the naked eye, while the liquid must be viewed under a black light.

The particle inspection process works by magnetizing a section of the casting. The test medium is then applied to the magnetized area. Any crack on, or near, the surface will interrupt the magnetic field. This causes the test medium to accumulate along the crack making it visible, figure 2-6. Be aware, for a crack to attract the test medium and become visible, it must cross the magnetic field. The test medium does not

accumulate along a crack that runs parallel to the magnetic field. To eliminate this possibility, rotate the magnet 90 degrees and repeat the test on the same area.

Pressure testing

Pressure testing can be used for both ferrous (cast-iron) and non-ferrous metals (aluminum). The test is performed by sealing off all of the coolant passages, then charging the water jackets with compressed air. Cracks are revealed by escaping air. Two types of equipment—universal pressure testers and submersible pressure testers—are available. Both methods will reveal surface cracks, as well as internal cracks, that are not readily visible.

With a universal tester, the head is mounted to a bench and adjustable pads are used to seal the passages. Compressed air is supplied through a coupling installed into one of the pads. A detergent solution is sprayed on the casting and any escaping air will cause the solution to bubble, figure 2-7.

Submersible testers require a variety of sealing plates to close off the coolant passages. An air hose is connected to one of the plates, and the head is lowered into a water tank. Cracks are revealed by a series of bubbles that rise to the surface of the water.

Dye penetrant testing

Dye penetrant testing is a chemical technique that will work for both ferrous and non-ferrous metals. This process reveals cracks only on visible surfaces of the casting. Different manufacturers have somewhat different procedures. Most use a dye penetrant, a dye remover, and a developing agent. Follow these general steps:

1. Clean the surface to be checked with a non-residual solvent.
2. Apply the dye penetrant to the surface, figure 2-8, and allow it to soak for several minutes.
3. Apply the remover and wipe both the remover and the excess penetrant from the surface.
4. Apply the developer and allow it to dry, figure 2-9.

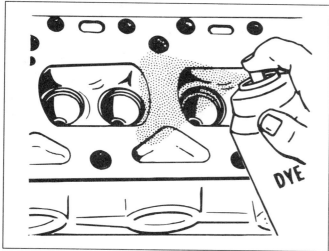

2-8 *Apply the dye penetrant and allow it to soak in.*

As the developer dries, it draws the dye penetrant out of the cracks and reacts with it to make it visible, figure 2-10.

VALVE SPRING INSPECTION

Give valve springs a thorough visual inspection. Look for nicks, pitting, corrosion, and cracks that might cause the spring to break while in service. Once the springs pass a visual inspection, check them for:

- Free height
- Squareness
- Tension.

Free Height Comparison

Arrange the springs in a line and lay a straightedge across their tops. Look for variations in length. All of the springs should stand at approximately the same height. Replace any that are not within $\frac{1}{16}$ inch of the others.

2-9 *Clean the surface, then apply the developer.*

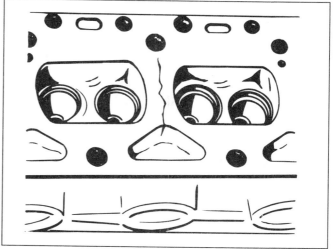

2-10 *As the developer dries, dye remaining in a crack becomes visible.*

Checking for Squareness

To test for squareness, place a spring alongside a square, then rotate the spring, figure 2-11. Squareness should not vary by more than $\frac{1}{32}$ inch at any location along the circumference of the spring.

Testing Tension

Spring tension is measured with a special gauge, figure 2-12. Specifications are generally provided for the two extremes—open pressure and seat pressure—at which the spring operates. Open pressure indicates spring tension, with the spring compressed and the valve fully open. Seat pressure indicates spring tension, with the spring at its installed height and the valve resting on its seat.

To test a spring, place it on the table, pull the lever down to the specified height, and observe the indicator. If tension is not within 90 percent of specifications at each length, discard the spring and install a new one.

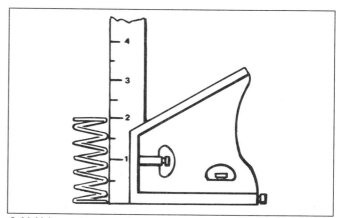

2-11 *Valve springs must be square and of equal height.*

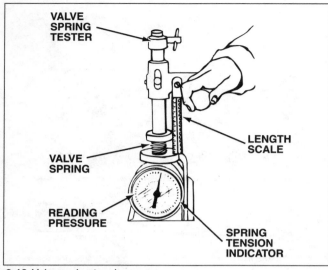

2-12 *Valve spring tension gauge.*

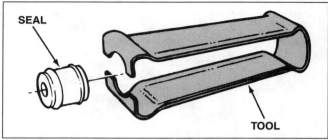

2-13 *Positive-lock valve seal installation tool.*

VALVE SPRING KEEPERS AND RETAINERS

Check fit each keeper onto the valve stem. If a keeper does not fit securely to the valve, or you notice any scoring, pitting, or other signs of wear, replace it. Look over the lock grooves machined into the valve stem. The groove must be clean, smooth, and well defined. Replace the valve if any sign of deformation is present.

Inspect shims, retainers, seats, and any other parts for signs of wear. If valve rotators are used, be sure they allow free movement without binding.

VALVE STEM SEAL SERVICE

Valve stem seals are installed to minimize oil flow down the valve guide and into the combustion chamber. Seal failure will increase oil consumption and cause heavy deposits to form on the back side of the valve. Valve stem seals will be one of three designs:

- O-ring
- Umbrella
- Positive lock.

Stem seals are routinely replaced when servicing the valves and can also be fitted without removing the cylinder head from the engine.

On-vehicle Seal Replacement

Special tools are required to replace valve stem seals without removing the cylinder head. To replace the seal, the valve spring must be removed. The valve must be held firmly against its seat, in order to do this. This can be accomplished by charging the cylinder with compressed air, while both the intake and exhaust valves are closed.

To perform the operation, position the piston in the cylinder to be serviced at TDC on the firing stroke. Insert an air fitting into the spark plug hole and attach an air line to it. The air pressure will hold the valve against its seat, so that keepers and spring can be removed without the valve falling into the cylinder. The piston must be exactly at TDC, or the air pressure will rotate the crankshaft.

Once the cylinder is pressurized, a special valve spring compressor is used to remove the valve spring. Remove and replace the seal. Then, reinstall the valve spring, retainer, and keepers.

On-bench Seal Replacement

O-ring and umbrella seals fit snugly to the valve stem. To install, lubricate the seal and slide it over the valve stem. Umbrella seals are fitted before the valve spring is installed. O-ring seals are installed after the valve spring, while the spring is compressed.

Positive-lock seals seat on the outside diameter of the valve guide, and must be installed before the valve is fitted. Use a seal installation tool, figure 2-13, to press the seal onto the valve guide. Then, lubricate the valve stem, fit the valve into the guide, and carefully push it through the seal.

VALVE GUIDE INSPECTION AND SERVICE

Valve guides may be either the **integral** or insert type. Integral guides are machined directly into the cylinder head casting, while insert guides are a separate piece that is press-fit into the head casting. Aluminum cylinder heads always have insert-type valve guides. Valve guides are wear items that demand attention whenever cylinder head repairs are performed.

Integral: A part that is formed into a casting.

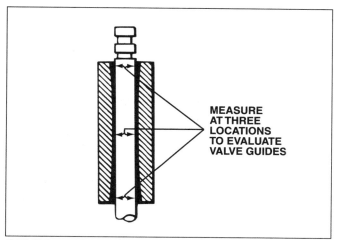

2-14 *Valve guide wear is greatest at each end; take three measurements.*

Residual pockets of carbon or gum often remain in the guides after the cylinder head has been cleaned. Therefore, it is important to clean the guides before measuring them. Use brushes and scrapers designed specifically for valve guide cleaning.

Measuring Valve Guides

To evaluate valve guide condition, two measurements are required: measure valve guide height and stem-to-guide clearance.

Valve guide height

Valve guide height is the distance the guide protrudes from the valve spring side of the cylinder head. Valve guide height can be measured with a machinist's scale.

Stem-to-guide clearance

Valve guides wear to a bell-mouth shape at each end, figure 2-14. Measure the inside diameter at the top, bottom, and center of the guide to get an accurate picture of guide condition.

There must be a minimal amount of clearance, generally 0.001 to 0.002 inch, between a valve guide and the valve stem. To determine this clearance, measure the inside diameter of the guide and the outside diameter of the valve stem. The difference between these two readings is the clearance. Measure guide inside diameter using a:

• Valve guide dial bore gauge
• Small hole gauge and outside micrometer
• Dial indicator.

The dial bore gauge will give the most accurate results. A fixture is used to zero the gauge to the valve stem diameter. Insert the calibrated gauge into the valve guide and clearance will register on the dial face.

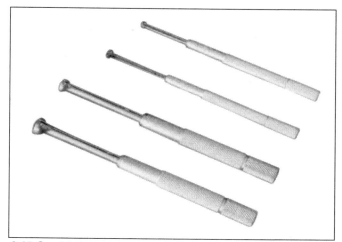

2-15 *Small-hole gauges are available in a number of sizes.*

To use a small hole or split-ball gauge, insert the gauge into the valve guide, figure 2-15. Adjust the thumb screw, so the fingers contact the guide, then lock the tool to hold the setting. Remove the tool and use an outside micrometer to measure across the gauge, figure 2-16. Subtract valve stem diameter from guide inside diameter to calculate clearance.

Using a dial indicator is the least accurate method of checking guide clearance. Place a new valve in the guide and attach a dial indicator to the cylinder head, so the indicator plunger touches the edge of the valve. Rock the valve to get a reading on the dial indicator. A new valve must be used to eliminate recording wear on the valve stem.

Integral Valve Guide Repair

Worn integral valve guides can be repaired with:

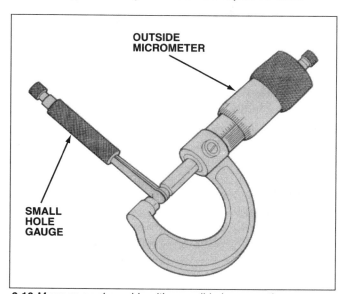

2-16 *Measure a valve guide with a small-hole gauge, then measure the gauge with an outside micrometer.*

- Valves with oversized stems
- Knurling
- Thin-wall liners
- False guides
- Coil wire inserts.

Oversized valve stems

Valves with oversized stems are of limited availability. To fit the valves, the inside diameter of the valve guide is bored to accept the larger diameter valve stem.

Knurling

Knurling is a repair that can be used only if the excess valve guide clearance is less than 0.005 inch. A special threaded arbor is fed through the guide to knurl it. The threads of the arbor push their way into the guide forcing metal out into the bore to create a smaller inside diameter.

After knurling, a valve guide **reamer** is used to bring the guide to final size. Final size should provide one-half the minimum specified clearance. The small clearance is necessary to provide as much service life as possible.

Thin-walling

Installing thin-wall liners is a method of restoring worn valve guides back to standard diameter. Most liners are made of bronze and have a wall diameter of about 0.010 inch. Thin-wall liners are used when guide wear is within 0.020 inch of specification.

The original valve guide bore is opened up with a piloting drill. The liner is pressed or driven into the bore, then the top is cut flush with the head using a special trimming tool. An **expander broach** is run through the insert to firmly seat it in the head. The inside of the insert is reamed to final size.

False guides

Integral valve guides worn beyond 0.020 inch can be repaired by installing a "false guide." False guides resemble the guide inserts used with aluminum heads and can be made of either cast-iron or bronze.

When installing false guides, a special fixture is used to align the drill bit and retain the centerline of the original guide. The bore is drilled slightly smaller than the outside diameter of the new guide, then opened with a reamer to obtain an interference fit. The false guide is pressed, or driven into the head, and the inside diameter is reamed to size.

Coil wire inserts

Coil wire inserts resemble the helical inserts used to repair damaged threads. The installed insert has a spiralled internal finish. The spiral helps retain oil, so valves can be fitted with about one-half the clearance required by solid wall guides. Installation procedures are similar to those used for thread repair.

Insert Valve Guide Replacement

All aluminum cylinder heads use valve guide inserts. These are serviceable items and should always be replaced whenever valve work is performed.

Valve guide inserts install with an interference fit, usually about 0.001 to 0.002 inch. Preheating the head will make removal and installation easier. Remove inserts using a valve guide driver. Measure the bore to determine the proper size for the new guide. Some standard replacement inserts are as much as 0.008 inch oversized, and the bore may have to be opened up to accept them.

Install the new insert into the head using the guide driver. Measure the height of the installed guide. It must have the exact amount of protrusion as the original. Ream the inside diameter to obtain the proper oil clearance.

VALVE INSPECTION AND SERVICE

Remove any carbon deposits from the valves by using a bench grinder with a soft-wire wheel. Perform a visual inspection to check for:

- Burning
- Guttering or channeling
- Necked valve stem
- Cracks
- Valve face wear
- Valve stem wear.

Inspecting for Wear and Damage

Burning, guttering, and channeling, figure 2-17, are caused by excessive valve temperatures. This condition is a result of preignition, poor seating, deposit accumulation, or metal erosion.

Valves do not always burn through from overheating. Sometimes the heat-softened valve will deform as it is forced into its seat by the springs and combustion

---

Reamer: A side cutting tool used to finish a drilled hole to an exact size.

Expander Broach: A tool used to seat a bushing and form the outside diameter of the bushing to the irregular surface of the bore.

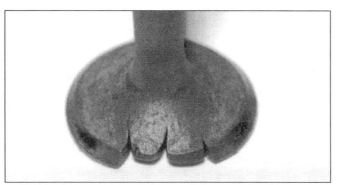

2-17 *Valve burning and guttering results from overheating.*

2-19 *Uneven cooling causes hoop stressing, which can cause a valve to crack.*

pressure. The head of the valve will pull down in the center to form a cup.

Necking, figure 2-18, occurs when the head of the valve pulls away from the valve stem, stretching the metal just above the head and causing it to thin. Necking is caused by overheating or too much valve spring pressure.

Cracks often result from "hoop stressing," a common overheating failure. Valves exposed to temperatures higher than they are designed to withstand cool unevenly. The outer edge of the valve cools quickly and contracts around the hot center. This temperature differential can stress the valve to the breaking point, figure 2-19.

Look the valve face over for signs of wear. Normal operating conditions will wear an even groove around the middle of the valve face, where it makes seat

contact. Check for signs of hot spots, peening from debris, and recession caused by high seating forces.

Check the valve stems for signs of galling, scoring, and binding. High seating forces and uneven seat pressure can develop stress cracks in the stem just above the valve head, figure 2-20. Excessive valve lash and guide clearance, worn rocker arms, and weak springs can cause wear at the tip of the stem near the keeper seats. Check the valve tip for mushrooming and other deformities.

Next, measure the valves to ensure they are in usable condition. Take three micrometer readings along the valve stem to check for wear and taper. Overall stem diameter must be within factory specifications. Taper should not exceed 0.001 inch. Check for

2-18 *Necking refers to a narrow area on the valve stem caused by the stretching.*

2-20 *Check for scoring or galling on the valve stem and for signs of stress at the tip end.*

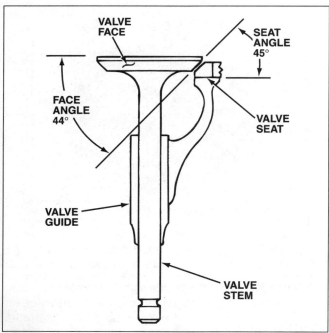

2-21 *An interference angle, different angles on the valve face and seat, improves valve seating on initial startup.*

bent stems using V-blocks and a dial indicator. Use a vernier caliper to verify **valve margin**. There must be at least ¹⁄₃₂ inch of margin after grinding the valve face.

Grinding Valve Faces

The valve face is reconditioned using a valve grinding machine. Procedures will vary slightly between types of equipment. However, the basic operations are the same regardless of the valve grinder being used.

Most valves are finished at a 45-degree angle, although some intake valve faces are ground to 30 degrees. Some manufacturers call for an **interference angle**, figure 2-21. This technique promotes positive seating on initial startup, improves valve sealing at high engine speeds and temperatures, and also helps to prevent carbon build-up. Although the interference angle can be machined into either the valve face or the seat, most machinists prefer to grind interference into the valve. A half degree of interference is generally sufficient, but common practice is to provide 1 degree of interference. For 1 degree of interference, the valve face is ground to 44 degrees and the seat to 45 degrees.

Keep these tips in mind when reconditioning valves:

- The chuck should grip the valve just above the wear area on the stem and as close as possible to the fillet to prevent wobble.
- Keep the grinding stone well dressed to ensure a good finish, and occasionally rotate the diamond tip to equalize wear when dressing.
- Switch on the spindle motor and visually check valve runout before taking a cut.
- Allow a constant stream of coolant to flow over the valve while grinding.
- Advance the stone feed wheel until the stone contacts the valve face, and begin stroking the valve across the stone immediately.
- Never allow any part of the valve face to pass beyond either edge of the stone while you stroke it.
- To get a clean, smooth, concentric face on the valve, advance the stone slowly. Never feed the stone in more than about 0.001 to 0.002 inch at a time. Wait for sparks to dissipate before advancing the stone further.
- Once the face cleans up, back off the stone feed. Never back the valve away from the stone. Moving the valve off of the stone will distort the edge of the valve.
- After grinding the valve, remeasure the margin and replace the valve if margin is less than 1/32 inch.

Grinding Valve Stems

Resurfacing, or "**tipping,**" the valve stem end provides a smooth, flat, square surface for the rocker arm or cam follower to contact. In addition, removing metal from the valve stem compensates for the valve sinking into the head as a result of valve and seat reconditioning.

With some valve grinders, the tip must be ground before the face to eliminate mushrooming and provide a true surface for the spindle chuck to grip. Tipping the valve to correct assembled height must be done after the valve seat has been reconditioned.

Usually, grinding 0.003 inch from the stem provides an adequate surface. To finish the valve, grind a 45-degree chamfer on the edge of the stem.

VALVE SEAT INSPECTION AND SERVICE

Valve seats can be either integral (cast into the head) or inserts (press-fit into the head); both types can be reconditioned and replaced.

Valve Margin: The distance on a valve from the top of the machined face to the edge of the valve.

Interference Angle: The difference between the angle at which the valve face is ground and the angle at which the valve seat is ground.

Tipping: Term for grinding the stem end of a valve to maintain correct stem height after grinding the valve face.

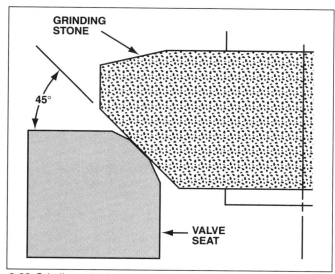

2-22 *Grinding a 45-degree angle establishes the valve seat.*

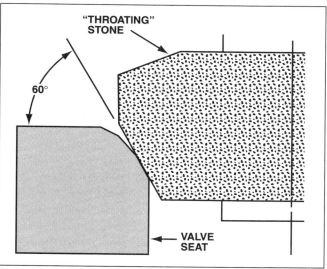

2-24 *Grinding a 60-degree angle removes metal from the bottom of the seat to raise and narrow the seat in the cylinder head.*

Inspecting for Wear and Damage

Most integral valve seats are induction-hardened to provide longer service life. A draw back to induction-hardening is that it creates stress in the casting and makes the valve seats prone to cracking. Inspect valve seats for cracks, burned areas, recession, and other damage. Minor damage can be repaired by grinding, or cutting new seating angles. A head with extensive seat damage can often be salvaged by boring out the old seat and installing an insert-type replacement seat.

Insert-type seats are inspected for cracks, burns, and other damage. Also, look for erosion on the cylinder head around the outside circumference of the seat. Check for looseness by gently prying up on the seat. If movement is detected, the seat must be replaced.

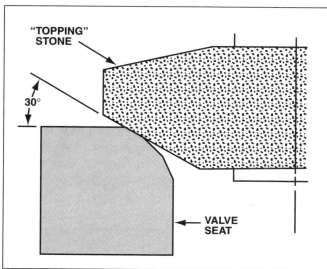

2-23 *Grinding a 30-degree angle removes metal from the top of the seat to lower and narrow the seat in the cylinder head.*

Seats that are solidly mounted in the head can be reconditioned.

Restoring Valve Seats

Three different angles must be established to recondition a valve seat. Most automotive valves use a 45-degree seating angle, figure 2-22. The seat width and position is corrected by grinding 30-degree, figure 2-23, and 60-degree angles, figure 2-24. For intake valve seats with a 30-degree face angle, use 15-degree and 45-degree correction angles to restore the seat.

All valve guide repairs must be performed before the seats are restored, because the valve guide is used to center the pilot for seat reconditioning equipment. If the guide is not in good condition, the seat will not be concentric to the guide, and the valve will not seal.

Seat grinding

The key to successful seat grinding is to select the proper stone and keep it well dressed to get a good surface finish. Stones are available in different harnesses for working different metals. Use a stone that is slightly larger in diameter than the seat but will not contact surrounding areas of the head.

To prepare for grinding, dress the stone, insert a pilot into the valve guide, and fit a lifting spring over the pilot. To grind a valve seat:

1. Grind the 45-degree angle to establish the seat. Work the stone into the seat with short, momentary contacts.
2. Remove the pilot and install the valve to check the fit. The seat must contact the valve face evenly along the entire circumference of the port.

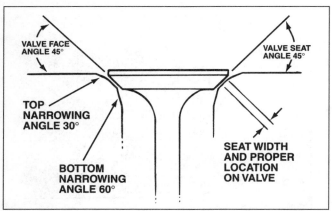

2-25 *Three-angle valve seat grinding positions the contact patch on the valve face for proper sealing.*

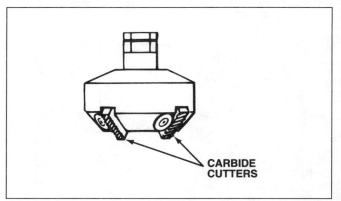

2-26 *Valve seats can be machined with special carbide cutting bits.*

3. Grind a 30-degree "**topping**" angle to lower the seat contact and position it on the valve face.
4. Grind a 60-degree "**throating**" angle to narrow the seat and move it up the valve face.
5. Recheck contact and position. If the contact patch is too wide, alternately top and throat until the desired width is achieved, figure 2-25.
6. If topping and throating makes the seat too narrow, correct by taking another light cut with the 45-degree stone.

Seat cutting

Seat cutting uses a carbide cutting bit, rather than a stone, to remove metal from the seat, figure 2-26. Cutting works well on bronze seat inserts that are difficult to grind and hard on stone life, soft seats that clog grinding stones to cause galling, and replacement seats that require excessive removal of metal. Another advantage of cutters is that they do not require dressing and always cut a true angle.

Most seat cutters form one angle at a time and operate similarly to seat grinders. Establish the seat, then correct the height and width by topping and throating. Single point cutting, a recent development, uses three blades on the same head to machine all three seat angles at once.

Replacing Valve Seats

Both integral and insert valve seats can be replaced if wear or damage is severe. For integral seats, a counterbore is cut into the old seat, then a new insert-type seat is press-fit into place. With insert seats, the old seat is removed and a new one is fitted.

When replacing integral seats with inserts, keep these points in mind:

- Bore a hole 0.005 to 0.008 inch smaller than the new seat insert to provide an interference fit.
- Bore to the exact depth of the new insert so the seat will fit flush to the head. Double check the depth before installing the insert.
- Chilling the seat in a refrigerator or freezer, or heating the cylinder head, will help overcome the interference and allow the seat to go in easier.
- To ensure that the seat does not come out, stake it in place after it is installed in the head.

There are several methods to remove original equipment insert seats. Some can be driven out with a long punch through the ports. Others can be lifted out with a pry bar. To remove difficult seats, weld a narrow bead around the inside of the seat. This will cause the seat to shrink and simplify its removal.

Valve-to-Seat Contact

Proper valve-to-seat contact is critical for efficient engine operation. The following checks are made during the machining process to ensure good contact: checking valve-to-seat contact and checking valve seat concentricity.

Checking valve-to-seat contact

A common practice for checking seat contact is to apply a light coating of Prussian blue to the valve face. Install the valve stem in the valve guide. Hold the valve about an inch above the seat, then gently lower it until the valve face contacts the valve seat. Do not rotate the valve while it is on the seat. Remove the valve and examine the impression on the valve face.

Topping: Term for lowering a valve seat in the cylinder head and narrowing the valve-to-seat contact patch by grinding an angle 15 degrees smaller than the seating angle.

Throating: Term for raising a valve seat in the cylinder head and narrowing the valve-to-seat contact patch by grinding an angle 15 degrees greater than the seating angle.

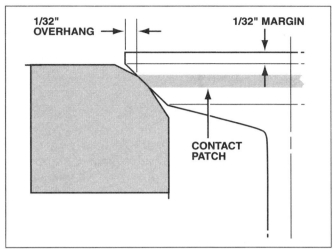

2-27 *A properly seated valve has an even seat contact patch with adequate overhang and margin.*

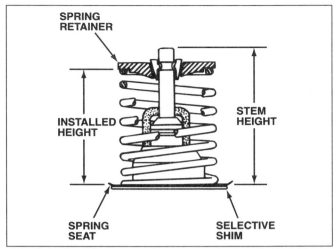

2-28 *Selective shims are used to correct valve installed height during assembly.*

The contact patch should be about $\frac{1}{16}$ to $\frac{3}{32}$ inch wide, with a minimum of $\frac{1}{32}$ inch overhang between the top of the seat and the edge of the valve face, figure 2-27. As a general rule, it is better to leave a wider seat on the exhaust valve because it runs hotter than the intake valve. Although a narrow seat provides better sealing, it also limits heat transfer.

Checking valve seat concentricity

Special dial indicating fixtures are available for checking valve seat **concentricity**. The tool fits a pilot into the valve guide and the plunger rests on the valve seat. The indicator is zeroed, then slowly rotated around the seat. The dial will register runout. Ideal runout is within 0.001 inch, although up to 0.002 inch is generally acceptable. Correct high spots by grinding or cutting.

VALVE AND VALVE SPRING ASSEMBLY

It is important to give the cylinder head a final cleaning before the valves are installed. All traces of abrasive dust and metal chips from machining must be removed. Any core plugs that fit to the head are installed at this time. Then, final checks and adjustments can be made.

Measuring Valve Stem Height

To measure valve stem height, fit the valve and hold it tightly to its seat. Then, measure the distance from the tip of the stem to the spring seat, figure 2-28. Compare this measurement to the one taken before reconditioning.

If the valve stem is too high, it can be corrected by tipping the valve as described earlier in this chapter. When valve stem height is too low, correct by grinding the valve seat.

Measuring Valve Spring Assembled Height

Spring assembled height, or installed height, is the distance from the spring seat to the bottom of the spring retainer. With the valve in place, install the retainer and keepers, without the spring, and measure with a telescoping gauge or machinist's scale.

If the assembled height is too great, shims can be placed under the spring to reduce it. Correct insufficient assembled height by grinding either the valve seat or the valve face.

Shim selection and installation

Valve spring shims are available in three sizes, 0.015, 0.030, and 0.060 inch, to correct installed height. Shims that are serrated on one side install with the serration facing down, toward the cylinder head.

Assembling Valves to the Cylinder Head

Generously lubricate the valve stem and guide with engine oil. If positive-lock seals are used, they must be installed before the valves are fitted. Slip the valve into the guide and hold it against its seat. Install umbrella-type valve guide seals before fitting the spring. Slide the seal down over the valve stem until it contacts the guide. Fit the shims, springs, and retainer, then

---

Concentricity: Having the same center, such as two circles drawn around a common center point. A valve guide and valve seat must be concentric, or centers at the same point.

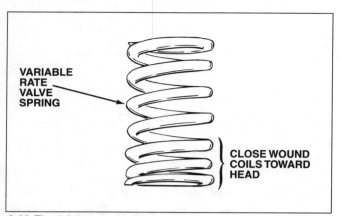

2-29 *The tightly wound coils of a variable-pitch valve spring install toward the cylinder head.*

compress the spring. O-ring seals are installed with the valve spring compressed. Fit the keepers securely into the groove on the valve stem. Slowly release spring tension and remove the valve spring compressor.

Some engines use **variable-pitch springs**. These have coils wound closer together at one end of the spring than at the other end. The tightly coiled end of a variable-pitch spring always installs toward the cylinder head, figure 2-29.

PUSHROD AND ROCKER SERVICE

Pushrods, rocker arms, and related parts are all wear items that must be closely inspected whenever they are removed from the engine. These items are non-serviceable items and are replaced rather than repaired.

Pushrod Service

Pushrods are cleaned, inspected for excessive wear, and checked for straightness. Wash pushrods with solvent and run a length of wire through the bore of hollow pushrods to dislodge any trapped debris.

Look over the ends of the pushrods. If you find nicks or grooves, a rough surface finish, any deformation, or signs of excessive wear, replace the part. Inspect the oil drill holes of hollow pushrods. If the oil hole is oval shaped, the pushrod is worn out and must be replaced. Look at the sides of the pushrod for signs of rubbing and scuffing and replace it if any metal has been rubbed away.

A quick and easy way to check the straightness of a pushrod is to roll it across a surface plate, thick piece of glass, or other perfectly smooth surface. Straight

pushrods will roll smoothly across the surface, while bent rods will tend to hop as they roll. Pushrod runout can also be checked using a special fixture or a dial indicator and V-blocks. As a general rule, runout should be less than 0.003 inch.

Stud-mounted Rocker Arm Assemblies

Wash the rocker arm parts in solvent and dry with compressed air to prepare them for inspection. Check rocker arms for wear on the valve stem contact area, the pushrod seat, and the pivot area. If valve stem wear is off-centered or shows grooves, pits, scores, or other irregularities, replace the rocker arm. Look for a ridge built up around the circumference of the pushrod seat. Also, check for galling, pitting, scoring, or indications of the pushrod hammering against the rocker arm. Inspect the sides of the rocker around the stud opening for stress cracks and uneven wear.

Inspect the face of pivot balls and fulcrums for scoring, galling, or pitting. Some pivots and fulcrums are relieved to direct oil flow. Replace the part if wear is to the bottom of any relief machined into the contact surface.

Shaft-mounted Rocker Arm Assemblies

The rocker shaft assembly on most engines can be unbolted and lifted off the engine as a unit, then disassembled on the bench. Always loosen and tighten the retaining bolts in small increments and in sequence. This will equalize valve spring pressure on the shaft to prevent bending the shaft.

Remove the cotter pin, roll pin, or any other fastener that holds the shaft assembly together. Then, slide the rocker arms, springs, spacers, and other parts off the shaft. Wash all components in solvent and lay them out in order on the work bench for inspection.

Inspect the shaft. The shaft surface should be smooth and free of any galling, scoring, or signs of excessive wear. Normal wear patterns will be greatest on the underside of the shaft, facing the cylinder head, due to spring pressure. Techniques previously detailed for inspecting stud-mounted rocker arms also apply to shaft-mounted arms. In addition, look over the shaft bore and replace the rocker if any deformation is found. Check for broken springs or any signs of bending, binding, or cracking. Spacers should spin freely on the shaft. Look for indications of binding, galling, or scoring on the internal bore and signs of undercutting on the end faces of the rockers.

Variable-pitch Spring: A spring that changes its rate of pressure increase as it is compressed. This is achieved by unequal spacing of the spring coils.

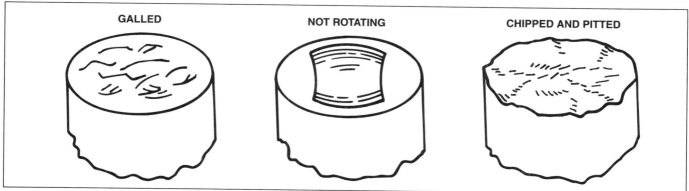

2-30 *Typical valve lifter wear patterns.*

VALVE LIFTER AND CAM FOLLOWER SERVICE

Valve lifters, also known as tappets or cam followers, are cylindrical parts that ride directly on the camshaft to open and close the valves. All engines use some type of lifter. Designs and service techniques will vary.

Overhead-valve Engine Lifter Service

Virtually all modern OHV automotive engines are fitted with hydraulic lifters. The hydraulic lifter uses engine oil pressure to take up slack in the valve train, so the valves operate with zero lash.

The pushrods must be removed to access the lifters. Because of the varnish build-up at the lower end of lifter bodies, you may have to soak the bores with penetrating oil to get them out. Once free of varnish, the valve lifters can be lifted from their bores by hand or with a special tool. Keep valve lifters in order, so they can be reinstalled into the same bore.

Inspect valve lifters for wear on the base. If wear is normal, the lifter base will be worn from a slightly convex surface to a concave surface. However, the wear pattern should never extend all the way to the edge of the lifter face. Pitting on the surface indicates excessive wear. Varnish build-up may cause valve lifters to stop rotating in their bores. Figure 2-30 shows typical wear patterns. The lifter base must have a slight crown. Check with a straightedge. If the straightedge lies flat, the lifter is worn out and must be replaced. Look over the sides of the lifter for signs of scoring and galling. Check the pushrod seat for indications of excessive wear and hammering.

Hydraulic lifters can be disassembled by removing the snapring holding the pushrod seat, figure 2-31. Remove the internal parts, wash in clean solvent, and inspect for wear and damage. Internal lifter parts are not available individually, and you must replace the entire lifter if any damage is found.

Clean the valve lifter bores thoroughly to ensure free valve lifter rotation. In some engines, it is possible to use a brake cylinder hone to deglaze the bores.

Hydraulic valve lifters should be leak-down tested before installation. Test procedures are detailed later in this chapter.

Fill hydraulic lifters with oil and purge any trapped air before installation. This can be done without special tools by submerging the valve lifters in motor oil and depressing the plunger several times. When you install the valve lifters, oil the bores and check for free rotation. Lifters should slip in easily and rotate by hand.

Overhead-cam Engine Lifter Service

Overhead cam engines use two basic cam follower designs: the bucket type that fits between the cam lobe and the valve stem, and the finger type that opens the valve

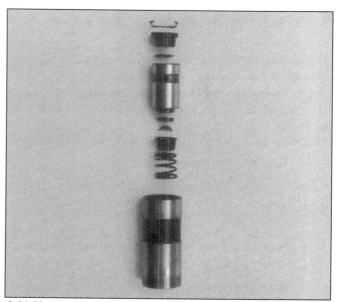

2-31 *Most hydraulic lifters can be disassembled for cleaning and inspection.*

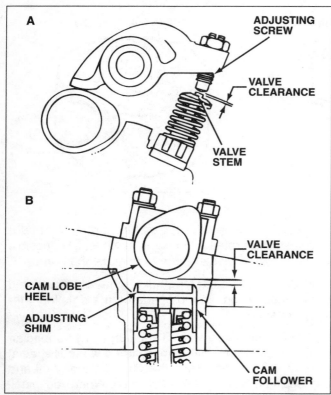

2-32 *Most OHC engines use either finger followers (A) or bucket followers (B).*

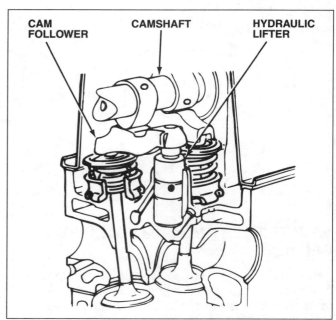

2-33 *Some OHC engines use finger followers that pivot on a remote hydraulic lifter.*

through lever action, figure 2-32. Valve lash with either type may be mechanically or hydraulically adjusted.

Bucket-type lifter removal and inspection

The camshaft must be removed to access bucket lifters. With the camshaft out of the way, simply lift the buckets out. Keep them in order, so they can be returned to the same location.

Inspect the lifters for signs of scoring and galling, and replace any defective units. Hydraulic bucket followers are not a serviceable item. Defective units must be replaced. Store hydraulic buckets upside down in a container of clean engine oil to prevent hydraulic bleed-down.

Finger-type lifter removal and inspection

Finger-type cam followers are similar in design to the rocker arms of an OHV engine. Often, they can be removed with the camshaft installed by relieving spring pressure on the valve. The lifter for most finger followers is a remote pivot that fits a bore machined into the head casting, figure 2-33.

Service procedures for finger-type hydraulic lifters are similar to those previously detailed for OHV lifters. With most designs, they can be disassembled, cleaned, inspected, and leak-down tested.

Hydraulic Lifter Testing

Hydraulic lifters, new and used, should be tested before installation. The best way to check lifters is with a leak-down test.

Leak-down testing

Leak-down testing checks the ability of a lifter to hold hydraulic pressure. A special tester is used, figure 2-34. The lifter is submerged in a test fluid, a weighted arm is placed on the pushrod seat, and the time it takes for the lifter to fully compress is recorded. Manufacturers provide leak-down rates for their lifters; normal leak-down time ranges from 20 to 90 seconds.

Kick-back testing

A kick-back test checks for internal binding. Hold the lifter upright and press down on the pushrod seat with your finger. Quickly release the lifter, and it should snap back into a fully extended position immediately.

VALVE LASH ADJUSTMENT

Typical valve lash clearance specifications range from 0.004 inch to 0.030 inch. Exhaust valves generally run with slightly more clearance than the intake valves. Use a flat feeler gauge to measure valve clearance. Make sure the gauge is clean and accurate. There are three common valve adjustment mechanisms:

- An adjustment nut holding the rocker to the rocker stud
- An adjustment screw on the rocker arm
- Selectively sized adjustment shims.

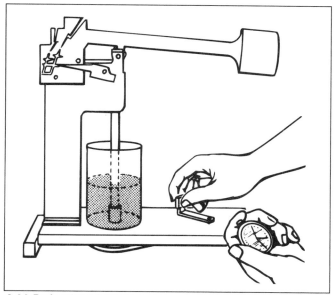

2-34 *Performing a leak-down test on a hydraulic lifter.*

For more information on valve adjustment procedures, refer to Book Eight of this series.

CAMSHAFT DRIVE SERVICE

Both OHC and OHV engines drive the camshaft through a gear set, a chain and sprockets, or a belt and sprockets. The camshaft drive is accessible for repairs on the front of the engine. Gear-to-gear setups are primarily used to drive the camshaft in heavy-duty OHV engines. Timing chains are common in OHV engines and are also used on some OHC engines. Timing belts are found on many OHC engines.

Disassembly, Inspection, and Measurement

To access the timing cover, the belts, fan, radiator, water pump, alternator, air-conditioning compressor, crankshaft pulley, harmonic balancer, and other items that attach to the front of the engine may need to be removed. Gear and chain systems will have either a stamped steel or a cast aluminum timing cover. Timing belts often have a lightweight plastic cover.

Remove stamped steel covers by removing the bolts and then gently prying the cover away from the block. Cast aluminum covers often bolt to the front of the block and to the oil pan and cylinder head as well. Most of the oil pan bolts may need to be loosened to lower the pan and allow the timing cover to be removed. Belt covers usually unbolt and remove easily.

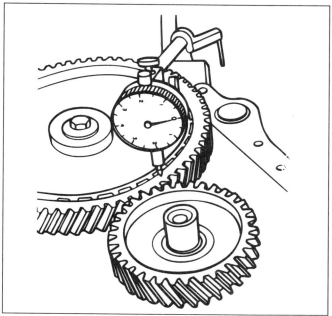

2-35 *Measuring camshaft gear runout.*

With the cover removed, rotate the crankshaft so that all of the timing marks are correctly aligned. The valve spring load on the camshaft must be removed to get accurate measurements. Remove the valve cover and loosen the rocker arms, or lash adjusters.

OHV gear drive systems

Camshaft and crankshaft timing gears might be pressed onto their shafts and require a puller to remove, although most are a slip fit. Before removing the gears, check backlash, runout, and camshaft endplay.

Measuring runout

To measure **runout**:

1. Set up and zero a dial indicator with the plunger resting on the face of the gear just inside the teeth, figure 2-35.
2. While watching the dial indicator, rotate the crankshaft through one complete revolution. Turn the crankshaft by hand; motion should be smooth and slow.
3. Note the maximum dial indicator readings, both positive and negative, while the gear is turning.
4. Add the two maximum dial indicator readings together to determine total gear runout.

Compare the runout to specifications for the engine being checked. Typical tolerance will be 0.004 to 0.005 inch for the camshaft gear and 0.003 to 0.005 inch for the crankshaft gear. If there is too much runout, check

---

Runout: Side-to-side deviation in the movement of a rotating assembly.

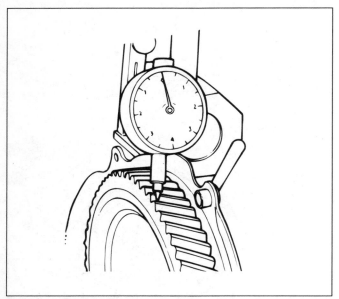

2-36 *Using a dial indicator to check timing gear backlash.*

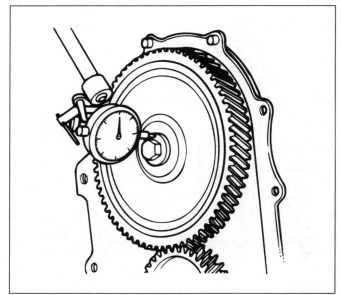

2-37 *Measuring camshaft endplay.*

for debris caught between the gear and the shoulder of the shaft behind it. Excessive runout can also be caused by a bent gear or shaft.

Measuring backlash

To measure **backlash**:

1. Attach a dial indicator to the engine so that its plunger tip is in line with the edge of the gear and resting on the edge of a gear tooth, figure 2-36.
2. Take all the backlash out of the gear mesh by rotating the crankshaft slightly. Then, zero the dial indicator.
3. Slowly rotate the crankshaft in the opposite direction just enough to take up all free play.
4. Note and record total backlash as displayed on the dial indicator.
5. Repeat this procedure on at least six gear teeth equally spaced around the camshaft gear.

Average the results of the six readings to get an accurate picture of gear backlash. Compare the results to backlash specifications listed in the service manual. Typically, specifications will be between 0.002 and 0.006 inch. If there is too much backlash, replace the gears.

Backlash can also be measured using a feeler gauge. To use a feeler gauge, take the backlash out of the gear mesh and insert a feeler gauge between the gear teeth of the slack side. Backlash will be equivalent to the thickness of the largest feeler gauge blade that will fit between the gears. Be aware, feeler gauge readings will not be as accurate as with a dial indicator.

Measuring camshaft endplay

To measure camshaft **endplay**:

1. Attach a dial indicator to the engine so its plunger is resting directly on the end of the fixing bolt, or on a machined surface near the center point of the gear, figure 2-37.
2. Push the camshaft toward the rear of the engine to take up all the clearance, then zero the dial indicator.
3. Insert two screwdrivers or pry bars under opposite sides of the gear. Gently pry out on the gear, then release tension on the screwdrivers.
4. The reading on the dial indicator reflects total camshaft endplay.

Compare findings to specifications from the manufacturer. Acceptable endplay is generally between 0.001 to 0.007 inch. If the endplay is outside limits, check the camshaft spacers or thrustplate for wear or damage.

OHV chain drive systems

Before removing the timing chain and gears, check chain slack as follows:

1. Rotate the crankshaft, without turning the camshaft, counterclockwise just enough to take all slack out of the left side of the chain.

Backlash: A lack of mesh between two gears, resulting in a lag between when one gear moves and when it engages the other.

Endplay: Movement along, or parallel to, the centerline of a shaft. Also called end thrust or axial play.

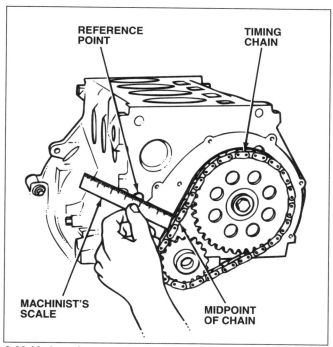

2-38 *Mark a reference point with all slack taken out of the chain.*

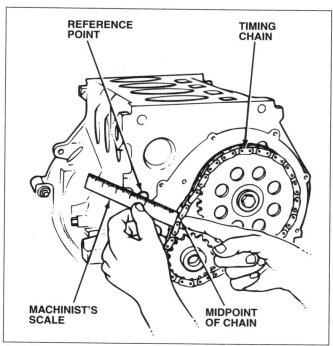

2-39 *Rotate to transfer slack to the opposite side, then measure chain deflection.*

2. Mark a reference point on the block face in line with the midpoint of the tensioned side of the chain. Then, measure from the reference point to the chain, figure 2-38.
3. Without turning the camshaft, rotate the crankshaft clockwise just enough to transfer the chain tension to the opposite side.
4. Push the slack portion of the chain toward the reference point and measure the distance from the reference point to the chain, figure 2-39.
5. Total chain deflection is the difference between the two measurements.

Compare your measurement to the specifications provided in the service manual. As a general rule, chain slack should be less than ½ inch.

To remove the chain and gear assembly, remove the bolt holding the camshaft gear and gently tap the edge of the gear to loosen it. The crankshaft sprocket is sometimes pressed in and must be removed with a puller. Slip both gears and the chain off of the engine.

Arrange the new timing chain and sprockets so that the timing marks align, and then simultaneously slip the sprockets over the camshaft and crankshaft snouts. Install the retaining bolts and draw the camshaft sprocket onto the camshaft by gradually tightening to specified torque. Check the camshaft endplay as previously described. Turn the crankshaft two complete revolutions in the normal direction of rotation, and check that the timing marks on both gears

line up again.

OHC chain drive systems

Chain driven OHC engines use a system similar to OHV engines. The main difference is that a chain tensioning device is used to take up slack in the long chain run. Chain tensioners, whether mechanical or hydraulic, are wear items normally replaced along with the chain. In addition, an intermediate gear may be used to drive the distributor or fuel pump. Be sure all timing marks, including those on the intermediate gear, are properly aligned.

OHC belt drive systems

With the timing cover removed, slowly rotate the crankshaft with a wrench as you inspect the entire length of the belt. Watch for:

- Dirt, coolant, and oil-soaked areas
- Hardened or cracked outer surface
- Separating cloth and rubber layers
- Worn, cracked, or missing teeth
- Wear or cracks on the side or back of the belt.

Also, inspect the crankshaft and camshaft sprockets, the tensioner, and any idler sprockets and pulleys for wear or damage.

To replace the belt, rotate the engine so all timing marks align. Loosen the tensioner and slip off the sprockets, figure 2-40. Check the tensioner assembly for wear and free movement. Slip the new belt over the

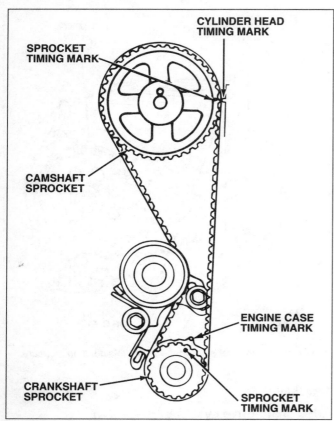

2-40 *Timing belt slack must be on the tensioner side when all of the timing marks are aligned.*

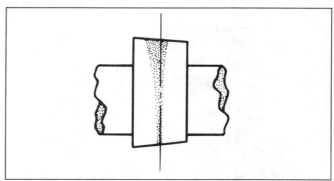

2-41 *Normal cam lobe wear creates a pattern that is wide on the nose and narrow on the heel.*

notches in the crankshaft sprocket, then fit it onto the camshaft sprocket. Keep the drive side of the belt as tight as possible. Turn the crankshaft slightly in the normal direction of rotation to take up any slack, then adjust the tensioner. Slowly turn the engine two complete revolutions and check to make sure all of the timing marks line up.

CAMSHAFT INSPECTION

The camshaft must be removed from the engine to check it for wear, damage, and straightness. When a new camshaft is installed, the lifters must also be replaced. Check camshaft:

- Surface finish and condition
- Straightness
- Journal diameter
- Lobe configuration and runout.

Inspecting for Wear and Damage

Look for wear patterns on the camshaft lobes. A normal wear pattern is wide on the nose, narrow on the

base, and will not extend to the edges of the lobe, figure 2-41. Pitting, scoring, or rounding of the camshaft lobes are immediate signs of extreme wear which will affect **valve duration** and lobe lift. Replace the camshaft if extreme wear is found.

Checking Camshaft Straightness

A dial indicator and a set of V-blocks are used to check camshaft straightness. Support the shaft in the V-blocks by the front and rear bearing journals. Position the dial indicator, so its plunger rests on the center bearing journal, and zero the gauge. Watch the dial indicator as you slowly rotate the shaft through one complete revolution. The difference between the highest and the lowest dial indicator readings is the total camshaft runout. As a general rule, runout should be less than 0.002 inch.

Measuring the Camshaft

Both the bearing journal diameters and the lobe heights are measured to determine overall camshaft condition. All measurements can be taken with an outside micrometer.

Journal diameter

Be aware, each camshaft bearing journal is a different diameter on an OHV engine. The smallest journal is at the rear of the engine, and the largest is at the front. This allows you to easily fit the shaft through the bearings when installing it into the engine block. On an OHC engine, the bearing journals are usually all the same size when the camshaft is held in place by bearing caps.

On any camshaft, journal diameter can be quickly and accurately measured with an outside micrometer. Measure each journal at least twice in different locations around the circumference. Also, measure at both ends of the journal to check for taper. Compare your

Valve Duration: The number of crankshaft degrees that a valve remains open.

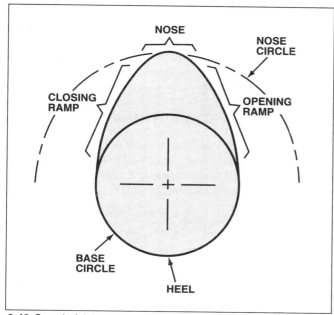

2-42 *Camshaft lobe nomenclature.*

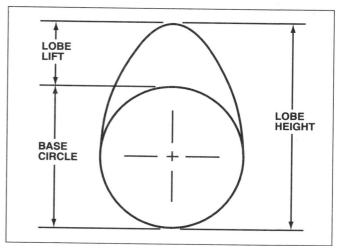

2-43 *Camshaft lobe lift is the lobe height minus the base circle diameter.*

findings to specifications to determine if the camshaft is within tolerance.

Base circle

Measure across the cam lobe to determine the **base circle** diameter. Measure the base circle diameter on all of the lobes; readings should be identical. An outside micrometer can be used for most camshafts. However, the ramps on some high-performance cams begin below the base circle center line. These cannot be measured with a micrometer.

Cam lobe height

Cam lobe height is the distance from the heel to the nose of the lobe, figure 2-42. Use an outside micrometer to measure the height of all the cam lobes, and take readings in the center and at both edges of the lobe. Expect to see up to about 0.002 inch of taper across the nose of a cam lobe. This can be considered normal and is machined into the lobe to help rotate the lifter in its bore.

Calculating camshaft lift

Camshaft lift is the distance that the cam lobes raise the lifters. Calculate lift by subtracting the base circle diameter from the lobe height, figure 2-43. Lift can also be measured with a dial indicator. Install the dial indicator so that its plunger rests on the heel of a lobe. Zero the dial face, then rotate the shaft until the plunger is on the nose. The total dial indicator reading is equal to the cam lobe lift.

Base circle runout

Base circle runout, or concentricity, is measured with a dial indicator while the camshaft is supported in V-blocks. Position the dial indicator with the plunger resting on the heel of a cam lobe and zero the dial face. Slowly rotate the camshaft to the point where the gauge reading is lowest. Rotate the shaft in the opposite direction, so the plunger crosses the heel, to reach the lowest reading on the other side. The difference from side-to-side is the total base circle runout. Typically, the base circle runout should not exceed 0.001 inch.

CAMSHAFT BEARING AND BORE SERVICE

Inspect the camshaft bearings for signs of excessive wear and damage. Measure the inside diameter of the bearings with an inside micrometer or dial bore gauge. Measure camshaft bearing journals with an outside micrometer. Measure in several positions across the face to determine taper, and check at several points around the circumference to determine out-of-round. Determine the clearance by subtracting the journal diameter from the bearing inside diameter.

Removing and Installing Camshaft Bearings

The camshaft bearings on OHV engines are press fit to the engine block and must by removed using a special tool, figure 2-44.

Base Circle: An imaginary circle drawn on the profile of a cam lobe that intersects the very bottom of the lobe. It is the lowest point on the cam lobe in relation to the valve train.

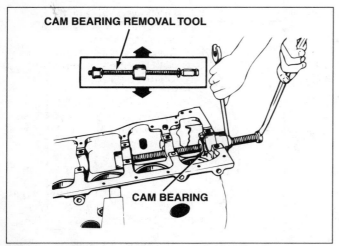

2-44 *Special tools are used to remove camshaft bearings.*

Aluminum OHC cylinder heads often do not have bearings, as the camshaft rides directly on journals machined into the head. However, split-shell camshaft bearing inserts, similar to those used for crankshaft main and connecting rod bearings, are used on some OHC designs. Other OHC engines use one-piece bearings that are press-fit into the head. These must be removed and installed with a forcing screw-type puller similar to those for an OHV engine.

New camshaft bearings are installed in an OHV engine using the same tool used to remove the old ones. Bearings must be checked for location because it is common to find different bearing diameters down the

length of the camshaft. Generally, large bearings are to the front and small bearings to the rear. Also, check the bearings for alignment so that the oil holes in the bearings will line up with the oil holes in the block. Install the bearings dry and fit them in order, working from the back of the engine toward the front.

CAMSHAFT INSTALLATION

Double-check the bearing oil hole alignment, then lubricate the bearings with engine oil. Coat the camshaft lobes with special high-pressure camshaft lubricant before installing the camshaft.

Feed the camshaft slowly and carefully through the bearing bores. Avoid scraping the soft bearing faces with the cam lobes. Once installed, the camshaft should spin freely by hand. If binding occurs, remove the shaft and dress any burrs or high spots on the bearings with a bearing scraper. Install the camshaft thrust plate, if used. Also, install the core plug behind the rear cam bearing.

Checking Camshaft Timing

Check the camshaft endplay and install the camshaft drive gears, or sprockets and chain, as previously detailed. Remember, the position in which the camshaft drive is installed determines **valve timing**. This usually consists of aligning marks on the gears. Turn the engine two complete revolutions and check that the marks are once again in alignment. Lubricate the new chain, tensioner, and gears with engine oil before in-

Valve Timing: A method of coordinating camshaft rotation and crankshaft rotation, so that the valves open and close at the right times during each of the piston strokes.

1. Which of the following operations is **NOT** required when replacing a cylinder head gasket?
   a. Checking for warpage
   b. Checking for squareness
   c. Checking for leakage
   d. Cleaning the bolt holes

2. Cylinder head gasket failure may be due to all of the following **EXCEPT**:
   a. Coolant leaks
   b. Compression leaks
   c. Oil leaks
   d. Vacuum leaks

3. Which of the following should be checked before disassembling the cylinder head?
   a. Rocker arm alignment
   b. OHC cam bore alignment
   c. Valve spring installed height
   d. Surface warpage

4. Internal cracks on a cast-iron cylinder head can be located by which of the following methods?
   a. Magnetic particle testing
   b. Pressure testing
   c. Dye penetrant testing
   d. All of the above

5. Valve springs are bench tested for all of the following **EXCEPT**:
   a. Installed height
   b. Free height
   c. Tension pressure
   d. Squareness

6. Typical valve guide clearance would be:
   a. 0.010 to 0.020 inch
   b. 0.002 to 0.005 inch
   c. 0.005 to 0.010 inch
   d. 0.001 to 0.002 inch

7. The best repair for an integral valve guide that is worn 0.030 inch beyond specification would be:
   a. Knurling
   b. A thin-wall liner
   c. A false guide
   d. A coil wire insert

8. The most common valve seat angles are:
   a. 30° or 45°
   b. 15° or 30°
   c. 30° or 60°
   d. 45° or 60°

9. After a valve job, the spring assembled height is too great. It can be corrected by:
   a. Grinding the seat
   b. Installing new springs
   c. Installing shims
   d. Tipping the valve

10. Valve margin must be at least:
    a. ⅛ inch
    b. 1/16 inch
    c. 1/32 inch
    d. 1/64 inch

11. Umbrella-type valve guide seals are installed:
    a. Before the valve
    b. Before the spring
    c. With the spring compressed
    d. On to the guide

12. A concave face on the base of a hydraulic lifter is an indication of:
    a. Overheating
    b. Lack of rotation
    c. Lack of lubrication
    d. Normal wear

13. Which of the following items must be removed in order to remove hydraulic lifters from an OHV engine?
    a. The camshaft
    b. The rocker arm
    c. The pushrod
    d. The cylinder head

14. Excessive camshaft endplay can be caused by:
    a. Worn timing gears
    b. A worn thrust plate
    c. Debris behind the thrust plate
    d. Worn timing chain

15. Camshaft lift is the difference between:
    a. Base circle diameter and lobe height
    b. Journal diameter and lobe height
    c. Camshaft runout and base circle diameter
    d. Base circle runout and lobe height

Chapter Three

ENGINE BLOCK
DIAGNOSIS AND REPAIR

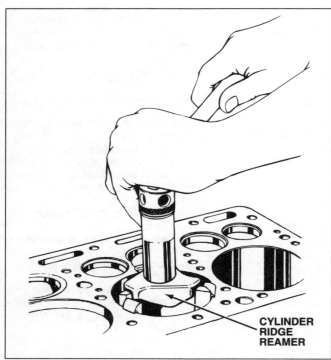

3-1 *Using a reamer to remove the cylinder ridge.*

Most engine block repairs require that the engine be removed from the vehicle, although some repairs, such as replacing crankshaft seals, oil pans, and timing covers, can be performed with the engine in the chassis. This chapter covers only the engine block operations performed on the bench.

ENGINE BLOCK DISASSEMBLY

To make disassembly easier and prevent overlooking hidden fasteners, steam clean or pressure wash the engine before you begin tearing it down. Watch for clues as to what caused the engine to fail as you disassemble it. Look for:

- Damaged or worn-out parts
- Asymmetric wear patterns
- Non-stock components and internal marks
- Mistakes made during previous engine work.

Remove all accessories, brackets, and mounts that bolt to the engine block. This includes the distributor, dipstick tube, and oil-sending unit. Remove the intake and exhaust manifolds, cylinder heads, and camshaft drive assembly as described in the previous chapter.

Remove the oil pan bolts and take the oil pan off; remove and discard the pan gasket. Remove the oil pick-up, oil pump, and pump drive. Look the engine over to be sure nothing is left in place that can interfere with removing the internal parts.

Marking Bearing Caps

Before the pistons and crankshaft can be removed, the bearing caps and connecting rods must be marked so they can be reassembled in proper order. The best way is to use a set of numbered punches. If that is unavailable, use a pin punch to mark the parts. Never scribe or file markings on the caps or rods. This creates a stress point that can result in engine failure. Stamp identification marks on the sides of bearing caps near the parting lines. Never stamp on the bottom of a bearing cap.

Main bearing caps are often numbered at the factory, but connecting rods and caps are not. Stamp the connecting rods and caps of an inline engine so the marks all face the same side of the engine. Markings on a V-type engine should all face the outside of the block.

Cylinder Wall Ridge Removal

The next step is to remove the ridge at the top of each cylinder bore to allow the pistons to come out without damage. A **ridge reamer** is used to cut away the ridge, figure 3-1. Turn the crankshaft to position the piston at

---

Ridge Reamer: A hand-operated cutting tool used to remove the wear ridge at the top of a cylinder bore.

---

bottom dead center (BDC), then fit the ridge reamer in the bore, and adjust the blade to take a light cut. Turn the reamer with a wrench to remove the ridge. Lead the cutting blade with the wrench by a few degrees. This helps smooth the cut and prevents undercutting or gouging. Clean all metal chips out of the cylinder after reaming the ridge.

Removing Piston Assemblies

Remove the bearing cap retaining nuts and lift off the cap. Slip protective covers or a small length of hose over the exposed ends of the rod bolts to prevent damaging the crank journal. Place the end of a hammer handle on the bottom of the piston and push the piston out of the bore. Remove both halves of the bearing shell. Keep the bearings in order for later inspection. Loosely reinstall the bearing cap onto the connecting rod.

Removing the Crankshaft

Remove the main bearing caps and lift the crankshaft free of the saddles. The flywheel may need to be removed first. However, on most engines, the flywheel can be removed along with the crankshaft. Set the crankshaft aside. Always store crankshafts in an upright position to prevent bending. Remove the main bearing inserts, keeping them in order for inspection. Then, reinstall and tighten the bolts to specified torque.

ENGINE BLOCK INSPECTION

Look the engine block casting over for obvious indications of damage. Carefully check around **core plugs** for signs of leakage, then remove the plugs. Any caps or plugs installed at the ends of the oil galleries must be removed as well. Check to make sure the block has been stripped down to a bare casting. Then, give it a thorough cleaning using one of the methods discussed in the previous chapter.

Water Passages and Oil Galleries

Once the block is clean and dry, run a rifle brush completely through the oil galleries to remove any residual sludge or trapped debris. After brushing, clear the passages with a blast of compressed air. Also, clean water jackets with a brush or scraper and compressed air to remove any scale or corrosion.

Crack Inspection

Carefully inspect the engine block for signs of damage. Check the block for casting cracks using magnetic particle inspection or a dye penetrant test. Both methods were discussed in chapter two of this book. Make note of any stripped or pulled threads that will need to be repaired before reassembly.

Deck Warpage and Surface Condition

Look the surface over for excessive scoring, corrosion, erosion, threads pulling up around bolt holes, cracks, dents, and scratches. Check for warpage using a straightedge and feeler gauge. Irregularities in the deck surface can usually be repaired by surface grinding.

REPAIRING DAMAGED THREADS

Inspect all of the threaded holes in the block for damage. Threads that are in good shape get cleaned with a thread chaser or bottoming tap as previously discussed for cylinder heads. Damaged internal threads can be repaired by:

- Drilling and tapping to oversize
- Installing a **helical insert**
- Installing an insert bushing.

Drilling and tapping to oversize can be used only in non-critical areas. A larger bolt installed in a critical area will not properly tighten to specified torque and can cause future problems.

Helical inserts, better known by their trademark name Helicoil®, restore threaded holes to their original size. When correctly installed, a helical insert is often stronger than the original threads, especially in aluminum castings.

Insert bushings are tubular, case-hardened, solid-steel wall pieces that are threaded inside and outside. The inner thread of the insert is sized to fit the original fastener of the hole to be repaired. Several types are available. All require the hole to be drilled considerably larger than the original, then tapped to the external thread size of the insert. Fasteners with damaged external threads should always be replaced rather than repaired.

Core Plug: A shallow, metal cup inserted into the engine block to seal holes left by manufacturing. Also called a freeze plug or expansion plug.

Helical Insert: A precision-formed coil of wire used to repair damaged threads or to reduce the internal diameter of a bored hole.

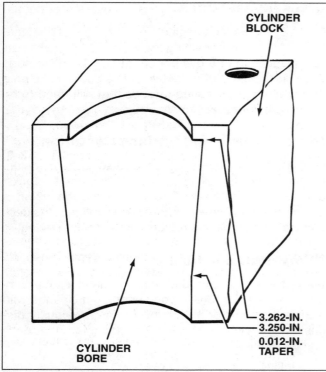

3-2 *Cylinder taper is the difference in diameter between the top and bottom of the bore.*

CYLINDER INSPECTION AND MEASUREMENT

Under normal conditions, a cylinder will develop two distinct wear patterns: taper and out-of-round. When the piston changes direction at the top of its stroke, combustion forces it into the cylinder wall. As a result, wear will be greatest near the top of the bore. This is known as taper, figure 3-2. Reciprocating motion causes the connecting rod to push the piston to the side, perpendicular to the rod, as it moves in the cylinder bore. This causes the cylinder to wear more on the sides than on the front and rear. Eventually, the cylinder bore wears to an oval shape, which is known as out-of-round.

Begin your inspection by checking the cylinder bores for major damage, such as cracks, gouges, scor-

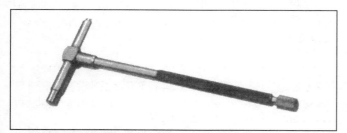

3-3 *A telescoping gauge along with an outside micrometer can be used to measure cylinder diameters.*

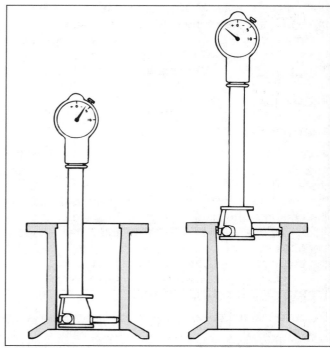

3-4 *Using a dial bore gauge to measure cylinder diameters and determine taper.*

ing, and broken cylinder walls. Next, measure the cylinder bore using precision instruments to evaluate wear damage.

Measuring Cylinders

Cylinders are measured using a dial bore gauge, an inside micrometer, or a telescoping gauge and an outside micrometer. A dial bore gauge gives the most accurate readings. A telescoping gauge is the least accurate tool for measuring a cylinder bore, figure 3-3. Regardless of what type of tool you are using, several measurements are required to determine cylinder condition.

Checking for overbore

The first check is to see whether the cylinders were bored oversize during a previous rebuild. Measure bore diameter near the bottom of the cylinders below the ring travel area. Take readings perpendicular to and parallel to the crankshaft. Both measurements should be the same when you are measuring in a non-wear area. Compare your findings with the specifications listed by the engine manufacturer.

Measuring for taper

Two measurements are taken to check for taper. Measure cylinder diameter near the top of the bore just below the ridge and at the bottom in the unworn portion of the cylinder, figure 3-4. Taper is equal to the difference between the two measurements. A realistic limit

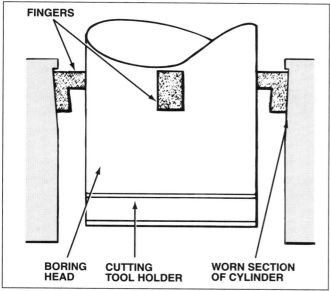

3-5 *Extendable fingers center the boring head in the cylinder.*

for cylinder wall taper on gasoline engines is about 0.005 inch. The limits for diesel engines are much lower, usually about 0.003 inch. To repair excessive taper requires boring to oversize or installing a sleeve in the cylinder.

Measuring for out-of-round

Determining out-of-round also requires two measurements. Measure cylinder diameter in a similar way as when checking for overbore; one measurement is taken in line with the crankshaft centerline, and the other is taken at a right angle to the crankshaft. However, these measurements are taken toward the top of the cylinder slightly below the ridge line. The difference between the two diameters equals cylinder out-of-round. A realistic limit for cylinder out-of-round, for both diesel and gasoline engines, is about 0.001 inch. Correcting an out-of-round cylinder requires machining.

CYLINDER WALL RECONDITIONING

If the cylinders are out of specifications for either taper or out-of-round, the bore should be opened up to a standard oversize. If cylinder wear is within specification, you can deglaze the cylinder walls and install new rings on the old pistons.

Cylinders can be oversized by either boring or honing. Both methods are discussed below. **Hypereutectic**

aluminum blocks must be honed to oversize. Attempting to bore a hypereutectic aluminum block can gouge the cylinder wall and cause irreparable damage.

Boring Cylinders

All boring equipment operates on the same principle. A single-point cutting bit clamps into a boring head at the bottom of a heavy, steel sleeve. The boring head is centered in the bore using extendable fingers, figure 3-5. The boring machine rotates the cutting bit, as it simultaneously feeds the boring head down through the cylinder to cut an oversized hole.

The surface in a freshly bored cylinder is too rough and irregular to seat piston rings correctly. Therefore, the tool bit is set to cut a diameter not quite as large as the final size. The cylinder is then brought out to the final size by honing to prepare the surface for new piston rings.

Boring can remove a considerable amount of metal with one pass. In order to bore a cylinder to 0.030 inch oversize, the boring head can be set to cut about 0.027 inch of metal from the wall. The remaining 0.003 inch of metal is removed by honing, which ensures a good surface finish for positive ring seating, figure 3-6.

Honing Cylinders

Honing machines use abrasive stones to remove metal from the cylinder bore. The stones mount in a honing head and are adjusted to cut an exact diameter. The head mounts to the machine and rotates as it moves up and down through the bore. Different grit stones are available for rough and finish cutting.

Cylinders can be oversized completely by honing. However, this is a slow procedure, and it shortens stone life. Most machinists prefer to bore the cylinders first, then hone them to final size.

Unlike boring, honing leaves a **plateau surface** that can support an oil film for the rings and piston skirts. Honing also establishes a **crosshatch** pattern on the cylinder walls, figure 3-7. The speed of the vertical and rotational movement of the honing head through the cylinder bore determines the angle of the crosshatch pattern. The crosshatch retains oil long enough to lap the piston rings to the cylinder walls, so they form a gas-tight seal.

Hypereutectic: A casting process that combines aluminum with small silicon particles. The silicon particles provide a durable surface finish.

Plateau Surface: A finish in which the highest points of a surface have been honed to flattened peaks.

Crosshatch: A multi-directional surface finish left on a cylinder wall after honing. The crosshatch finish retains oil to aid in piston ring seating.

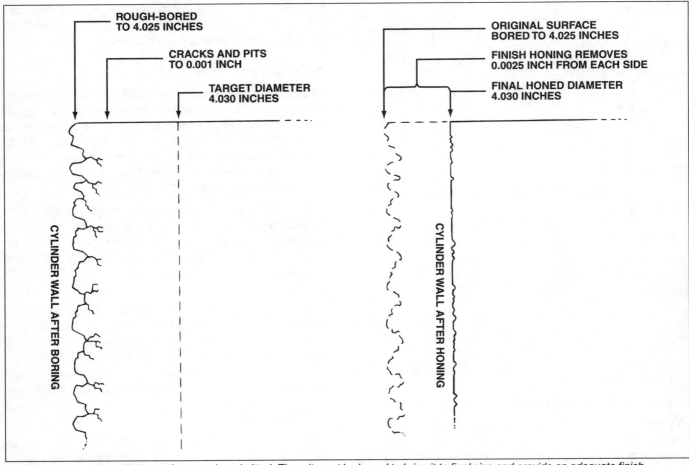

ROUGH-BORED TO 4.025 INCHES

CRACKS AND PITS TO 0.001 INCH

TARGET DIAMETER 4.030 INCHES

CYLINDER WALL AFTER BORING

ORIGINAL SURFACE BORED TO 4.025 INCHES

FINISH HONING REMOVES 0.0025 INCH FROM EACH SIDE

FINAL HONED DIAMETER 4.030 INCHES

CYLINDER WALL AFTER HONING

3-6 *Boring leaves the cylinder surface rough and pitted. Then, it must be honed to bring it to final size and provide an adequate finish.*

After the cylinders are brought out to final size by honing, the bore reconditioning process is finished by grinding a 45-degree **chamfer** on the top edge of the bore.

Deglazing Cylinder Walls

Cylinder deglazing is a cylinder wall reconditioning method that is used only when both the cylinder walls and the pistons are in good condition. Deglazing roughens the surface of the cylinders without significantly changing their overall diameter. The rough wall surface is necessary to break in new rings and allows the rings to conform exactly to the cylinders and form an effective seal.

A glaze breaker is used to deglaze the cylinders. The glaze breaker chucks into a drill motor and uses abrasive stones to prepare the cylinder walls, figure 3-8. The drill is operated at low speed, between 200 and 500 rpm, and stroked up and down in the cylinder bore. Stroking the tool creates the required crosshatch pattern on the cylinder wall. When deglazing cylinders,

lubricate the stones with a light-weight oil to keep particles in suspension and prevent heat build-up. Avoid using solvents or other chemical agents.

Cleaning Cylinder Walls

Wash freshly machined cylinders with hot soapy water. Never use solvents or any other type of cleanser as these leave a residue on the cylinder wall. Any trace of dirt left behind can cause hot spots and prevent ring seating. Contaminants in the engine turn the soapy water gray. Continue cleaning until the soap suds no longer change color. Use fresh, clean water to rinse off all the detergent. Dry the bores using compressed air and immediately spray the machined surfaces with a light lubricant to prevent corrosion.

CAMSHAFT BEARING SERVICE

Servicing the camshaft bearings was covered in detail in the previous chapter and will only be highlighted here.

Chamfer: A beveled edge.

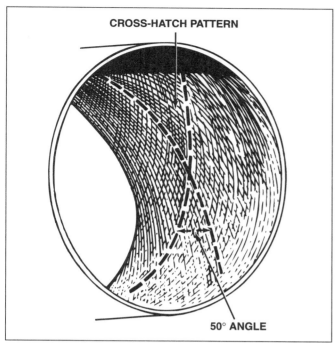

3-7 Proper honing technique leaves a crosshatch pattern on the cylinder wall.

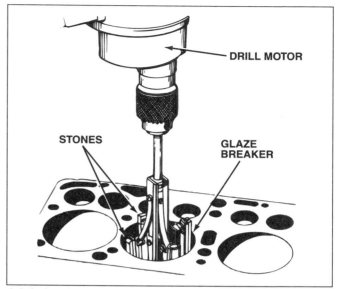

3-8 A glaze breaker deglazes cylinder walls to promote piston ring seating.

Visually inspect the bearings and bores for wear and damage. Measure the bores for out-of-round and misalignment. Damaged cam bearing bores on an overhead valve (OHV) engine can be line bored and fitted with oversized bearing inserts. Most overhead cam (OHC) engines fitted with camshaft bearing caps can be repaired back to standard size.

To restore OHC camshaft bearings to standard size, remove the bearing caps and grind a small amount of metal off the parting faces. This reduces the inside diameter of the bearing bores. Reinstall the bearing caps and open the bores to standard size by line boring.

CRANKSHAFT INSPECTION

Crankshaft inspection begins with a visual check for surface cracks and other damage. The reluctor ring for the engine speed sensor is an integral part of some crankshafts, and on some, it is a separate part that attaches to the crankshaft. With either design, carefully inspect the reluctor ring. Look for cracked, chipped, and missing teeth, as well as any other signs of damage. Once the crankshaft passes a visual inspection, it is checked for straightness, then the journals are measured for diameter, taper, and out-of-round. Bent crankshafts can be straightened by pressing, peening,

or heat stress-relieving. Damaged journals are restored to serviceable condition by grinding them to a standard undersize.

Crankshaft journals are generally reground to an undersize of 0.010 inch, 0.020 inch, or 0.030 inch. Oversized rod and main bearings are readily available in these increments for most engines. Normal practice is to grind all rod journals or all main journals to the same undersize, even when only one is damaged. Crankshaft journals that are within limits can be micropolished to improve the surface finish. **Micropolishing** simply restores the surface finish and will not correct for taper or out-of-round.

Surface Cracks and Damage

Look the shaft over for signs of cracking. Magnetic particle testing can be used to verify a crack. Cast-iron crankshafts usually self-destruct when a crack develops, so you will seldom find a cracked cast-iron crankshaft in a running engine. Forged-steel shafts are stronger, and you may occasionally remove a cracked one from a running engine.

Inspect the journal areas of the crankshaft. A slight discoloration is normal. Excessive discoloration, scoring, or pitting indicate problems. Check the snout for keyway damage and the flange for pilot-bearing bore damage and stripped threads.

Micropolishing: A machining technique that uses an abrasive belt to restore mild crankshaft journal damage while removing minimal amounts of metal.

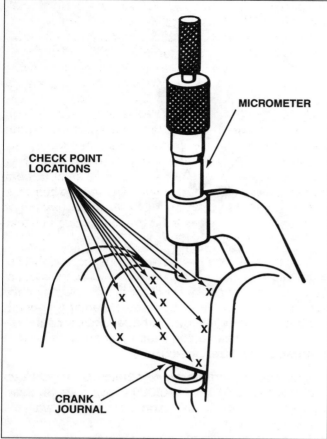

3-9 *Several outside micrometer measurements are taken to evaluate wear on a crankshaft journal.*

Check crankshaft straightness using a dial indicator and V-blocks. Position the indicator plunger on a main journal and rotate the shaft one complete revolution. Crankshaft bend equals one-half of total indicated runout.

Cleaning Oil Passages

If the crankshaft has threaded sludge trap plugs that seal the ends of the oil drillways, remove them to clean the oil passages. Wash the shaft with solvent and use a brush to clean the oil passages.

Measuring Crankshaft Journals

Bearing journals are measured with an outside micrometer. You will be checking for diameter, taper, and out-of-round. To get an accurate picture of wear, it is important to measure each journal at three locations: near the front and rear fillet radii and at the midpoint of the journal, figure 3-9. In addition, take at least two readings at each location. The second measurement should be taken 90 degrees around the circumference of the journal from the first.

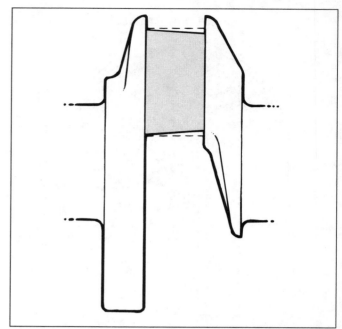

3-10 *Crankshaft journal wear with end-to-end taper.*

The maximum diameter reading is used to determine oil clearance and will reveal if the crankshaft was reground to undersize previously. Compare the readings taken 90 degrees apart to determine out-of-round. The three measurements across the journal determine taper. Three common bearing journal wear patterns are:

- End-to-end taper, figure 3-10
- Hourglass taper, figure 3-11
- Barrel taper, figure 3-12.

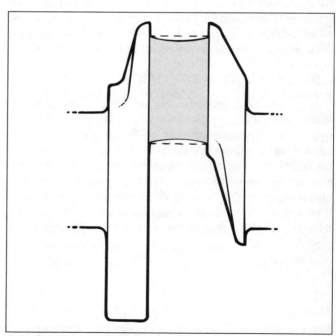

3-11 *Crankshaft journal wear with hourglass taper.*

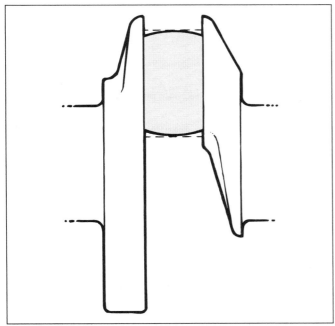

3-12 *Crankshaft journal wear with barrel taper.*

Compare your findings to specifications provided by the manufacturer. Crankshaft tolerances for taper and out-of-round are usually very tight. Modern engines often allow only 0.0002 to 0.0005 inch. In addition, any wear that reduces the diameter to 0.001 inch below standard requires the journal to be ground to an undersize.

MAIN AND CONNECTING ROD BEARING INSPECTION

Evaluating the main and rod bearing insert wear helps determine what areas of the block and crankshaft and which connecting rods will require close examination. Also, bearing wear patterns often provide insight as to what caused the engine to fail.

Evaluating Wear Patterns and Damage

Wipe the bearing shells clean with a rag and look them over. Line inserts up in pairs in the order they fit the engine. There should not be any scratches, embedded particles, or pieces of metal flaking off the bearing shells. The internal surface of the bearing shells should be uniformly gray and smooth. Wear should be greatest toward the center of the bearing and minimal near the **parting line** at the sides of the bore.

Look for asymmetric wear patterns on the bearings. Localized, smeared areas on the bearing indicate that

dirt particles were trapped between the bearing shell and the saddle. Flaking or wear at the edges of the shell can indicate that a shell is too wide for the journal, the bearing riding on the **fillet**, or a bent connecting rod. Connecting rod bearing wear at opposite sides and opposite ends of the two shells also indicates a bent rod. Main bearing wear concentrated on apparently random areas across the various bearings can indicate a warped block. Wear in a short arc on the bearing face indicates partial contact caused by excessive clearance. Even scoring across the bearing face indicates a poor finish on the crankshaft. A bent crankshaft is indicated by severe wear on the bearings at the center of the engine and a minimum amount of wear on the bearing shells that were installed at either end of the engine.

Turn the bearings shells over and inspect their backs. If the backs of the bearings show any scoring, unusual patterns, or appear highly polished, the bearings have spun in the bores. When this type of damage is found, the bearing bores will need to be reconditioned.

Calculating Oil Clearance

Proper bearing oil clearance is critical to engine service life. To calculate oil clearance, you need to know:

- Housing bore diameter
- Insert thickness
- Bearing inside diameter
- Journal outside diameter.

Housing bore diameter is the diameter across a bearing bore without the bearing inserts installed, figure 3-13. Measure housing bore diameter using either a dial bore gauge, inside micrometer, or telescoping gauge and outside micrometer.

Insert thickness is the true thickness of the bearing insert or shell. It is not the same as the nominal oversize or undersize. Oversize inserts have a larger outside diameter to fit an oversized bore, and undersize inserts have a smaller inside diameter to fit a reground journal, are both thicker than standard-sized bearings. A special outside micrometer with a ball-shaped anvil is used to measure bearing insert thickness. A standard micrometer will give a false reading because of the arch of the bearing.

Bearing inside diameter is the diameter across the hole for the journal, with the bearing inserts in place and the cap tightened to specified torque, figure 3-14.

Parting Line: The meeting points of two parts or machined pieces, such as the two halves of a split shell bearing.

Fillet: A curve of a specific radius machined into the edges of a crankshaft journal. The fillet provides additional strength between the journal and the crankshaft cheek.

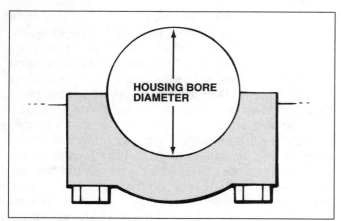

3-13 *Housing bore diameter is the distance across a bore without the bearings installed.*

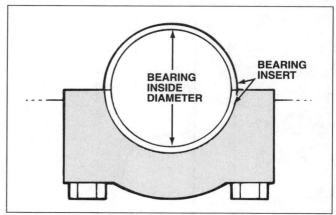

3-14 *Bearing inside diameter is the distance across a bore with the bearings installed.*

You can install the bearings and measure the inside diameter, or you can calculate it by adding twice the insert thickness to the housing bore diameter.

Crankshaft journal diameter is the maximum distance across the journal. Measuring crankshaft journal diameter was detailed previously in this chapter.

To determine the oil clearance, subtract the journal diameter from the bearing inside diameter. An alternative method for checking oil clearance is to measure it using Plastigauge®.

Plastigauge® is a string-like piece of plastic manufactured to a precise diameter. A strip of the plastic is placed between the bearing and the journal, and the cap is tightened to final torque, then carefully removed. Because the diameter of the gauging material is exact, tightening the cap crushes the Plastigauge® a specific amount. The amount of crush can be measured to determine oil clearance using a scale on the Plastigauge® package.

There are two important features to keep in mind when using Plastigauge®. First, the plastic material is not compatible with engine oil. Both bearing and journal must be perfectly dry. Second, never turn the crankshaft with Plastigauge® installed. This can smear the plastic and possibly damage the bearing.

Installing Main Bearings

Fit all of the bearing inserts to the block first, then to the caps. The block-side insert must have a drilled hole that aligns with the oil gallery opening in the saddle. Some manufacturers machine a relief into the upper shell that directs oil flow to the parting line at either side of the bearing. Be sure these bearing halves are installed into the engine block, not into the cap.

Bearing inserts assemble to the block and caps dry; do not lubricate them before you install them. Also,

wipe the backside of the bearing shells clean before you fit them. Install the bearing shells in order, working from one end of the block to the other, then fit the bearing shells to the caps. If the engine has separate thrust bearing inserts, fit them last. A couple of dabs of assembly lube on the outside edge of the thrust insert helps hold it in place. Avoid getting any lubricant behind the thrust bearings. Finish by wiping off the working surface of the bearings, then lubricate them with a thin film of engine oil.

Installing the Crankshaft

The crankshaft journals must be perfectly clean; wipe them down before you place the crank into the block. Keep the crankshaft parallel to the bearing bores as you slowly lower it straight down into position. Make sure the crankshaft does not gouge the soft surface of the thrust bearings as you lower it, and make sure that separate thrust bearing inserts remain fully seated as you install the crank.

Once in place, the crankshaft must fit squarely and be solidly supported by the main bearings in the saddles. The shaft must rotate freely and easily without resistance. If not, there is a problem. Binding may be caused by a bent shaft, misaligned bearing bores, incorrectly installed bearing shells, or defective or incorrectly sized bearings. Correct any problems now. Do not install the bearing caps if the crankshaft does not spin smoothly when resting in the block saddles.

Secure the crankshaft in the block by installing the main bearing caps using the following steps:

1. Fit the bearing caps over the journals and push down to mate them to the saddles. If you staggered the rear main seal parting line, add a drop of gasket sealer to the seal ends. Be sure to fit both ends of the seal precisely into the grooves.
2. Lubricate the bolt threads as required, then install the bolts and draw them up hand tight.

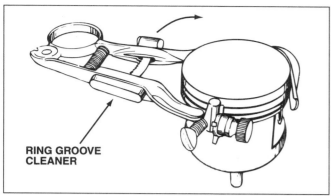

3-15 *A piston ring groove cleaner scrapes carbon, varnish, and other debris from the ring grooves of a piston.*

3. Rotate the crankshaft as you lightly tap the sides of each cap with a soft-face hammer to register it. This is an important step, as it squares the cap in the saddle.
4. Draw the bolts up as far as you can by hand.
5. Place a large screwdriver between a cap and crankshaft cheek and pry the crankshaft back and forth to align the thrust bearing. Do not pry on the cap that holds the thrust bearing.
6. Rotate the crankshaft at least one complete revolution to check for free movement.
7. Tighten the bolts in increments, alternating from side to side to bring them to final torque.

PISTON AND CONNECTING ROD SERVICE

Worn and damaged pistons cannot be repaired; they must be replaced. Connecting rods can generally be reconditioned and returned to service, unless they are severely damaged.

Cleaning and Measuring Pistons

Carefully remove any heavy deposits on the tops of the pistons with a gasket scraper. Do not scrape down to the metal because the piston tops are easily damaged. Clean remaining deposits by soaking the pistons in carburetor cleaner or washing them with solvent. Pistons also can be cleaned by bead blasting. However, bead blasting is abrasive and can damage the **ring lands** of a piston. Protect the ring lands by wrapping them with tape before bead blasting.

A critical operation is cleaning the piston ring grooves so that the new rings will slide freely when installed. For this job, special ring groove scrapers are available, figure 3-15. Some technicians use a broken segment of a piston ring, although care must be taken

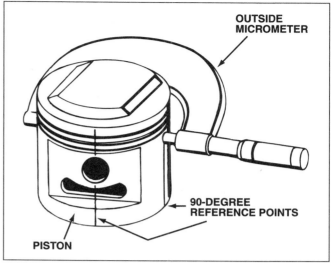

3-16 *Measuring piston diameter with an outside micrometer.*

not to shave metal from inside the grooves. Check the oil drain holes located at the back of the oil control ring groove to make sure that they are clear.

Inspecting and measuring pistons

Examine the entire piston, keeping a close eye out for unusual wear patterns on the skirts, ring lands, and wrist pin bosses. Inspect the wear patterns on the piston thrust faces for indications of a bent connecting rod. Wear patterns should be straight up and down, perpendicular to the wrist pin. An angled pattern indicates a bent rod. Check the ring lands. Replace the piston if you find any scuffing, hairline cracks, or chips. Check the wrist pin bosses for signs of cracking and metal fatigue on the inside of the piston.

Three measurements are taken to evaluate overall piston condition. They are:

- Piston diameter
- Piston skirt diameter
- Ring groove side clearance.

Measure the piston diameter using an outside micrometer. Diameter is generally measured on the thrust surface at the centerline of the pin bore and perpendicular, at a 90-degree angle, to the wrist pin, figure 3-16. Most automotive pistons will measure 0.001 to 0.003 inch less than the cylinder bore diameter.

Check for collapsed skirts by measuring across the thrust surfaces at the bottom of the skirt with an outside micrometer. Compare this figure to piston diameter. The piston should be about 0.0015 inch wider at the bottom of the skirt.

Ring Lands: The part of a piston between the ring grooves. The lands strengthen and support the ring grooves.

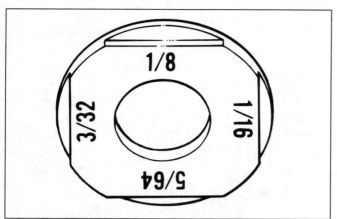

3-17 *A ring wear gauge measures the width of a piston ring groove.*

A new piston ring is required to calculate ring groove side clearance. Typical side clearance for automotive engines is in the 0.001 to 0.003 inch range. There are two methods of measuring ring groove side clearance. One uses a ring wear gauge and micrometer, and the other uses a blade-type feeler gauge. A ring wear gauge is a type of feeler gauge designed specifically to measure the ring groove width, figure 3-17. Measure the ring groove width with the wear gauge and measure the thickness of the new ring with an outside micrometer. Then, subtract the ring width from the groove width to calculate side clearance.

To measure side clearance with a feeler gauge, simply fit the new ring backward into the groove. Then, slip the feeler gauge blades between the ring and the

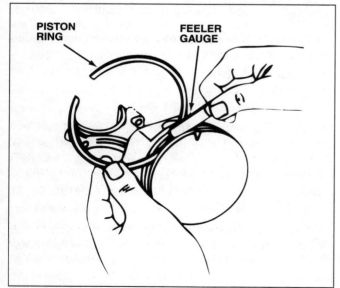

3-18 *Checking piston ring side clearance with a feeler gauge.*

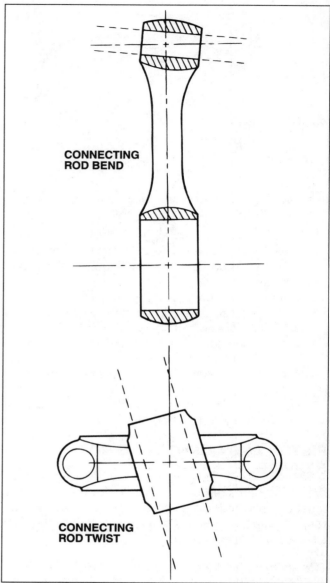

3-19 *Bend and twist are common connecting rod alignment problems.*

groove, figure 3-18. The largest feeler gauge blade that will fit between the ring and the side of the groove equals the side clearance.

Checking Connecting Rod Alignment

The connecting rod bore centerline at each end must be parallel to within approximately 0.001 inch for each 6 inches of connecting rod length, figure 3-19. If bore centerlines are out of parallel as viewed from the edge of the rod, the condition is called **bend**. If bore centerlines are out of parallel as viewed from the top, the

Bend: On a connecting rod, the condition of the two bores being out-of-parallel when viewed from the edge of the rod.

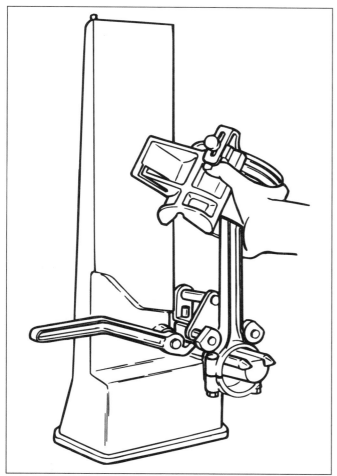

3-20 *A connecting rod alignment fixture quickly checks a rod for bend, twist, and offset.*

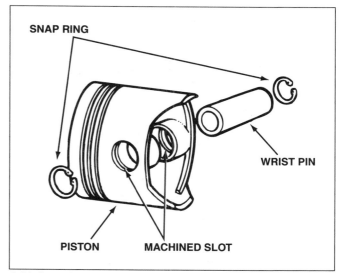

3-21 *Snap rings hold a full-floating wrist pin in the piston.*

condition is called **twist**. If the bore centerlines lie on different planes when the rod is viewed from its edge, the condition is called **offset**.

Connecting rod alignment can be checked using an alignment fixture, figure 3-20. Most fixtures allow you to check alignment either with or without the piston attached to the rod. Different adapters are used to check bend, twist, and offset.

Rod alignment problems can be corrected by cold bending the rod back into position using special tools. Damaged connecting rod bearing bores can be reconditioned by honing.

Separating Pistons and Connecting Rods

Two types of wrist pin, press-fit and full-floating, are commonly used on automotive engines. Press-fit pins

are installed with an interference fit of about 0.001 inch. Full-floating pins are installed without interference and are held in place by snap rings that fit into machined slots on the pin bore of the piston, figure 3-21.

Removing and measuring press-fit wrist pins

Press-fit wrist pins are removed with a press. Special adapters are used to support the piston and prevent damaging it while the pin is being removed. The pin is pressed out of the piston and rod with an arbor.

Once the pin is removed, the diameter of the wrist pin, the bore on the small-end of the connecting rod, and the pin bore on the piston are measured. Wrist pin diameter can be measured using an outside micrometer. The piston pin bore and connecting rod bore are measured using a special precision bore gauge, figure 3-22. The precision bore is generally an accessory item that attaches to a rod honing machine.

To measure pin bores with a precision bore gauge, it must be calibrated using a pin gauge setting fixture. Clamp two wrist pins into the setting fixture, figure 3-23. Then, fit the setting fixture onto the bore gauge and adjust the dial gauge to read the correct clearance, figure 3-24. Once the gauge is set, it can be used to check clearance in rod bushings or in piston bores.

When measurements are out of tolerance, oversized wrist pins can be installed. However, a considerable amount of machining is required, and it is more common to simply replace the piston and wrist pin with new ones.

Twist: On a connecting rod, the condition of the two bores being out-of-parallel when viewed from the top.

Offset: On a connecting rod, the condition of the two bores being out-of-parallel when viewed from the side.

3-22 *A precision bore gauge is used to measure piston pin and bore diameters.*

Installing press-fit wrist pins

Press-fit pins are installed using either a press or connecting rod heater. Press fitting is basically the reverse of removal, using an arbor and adapters. A connecting rod heater warms the small end of the connecting rod. This expands the pin bore and makes the pin fit easier. With either method, it is important to get the pin properly centered in the bores.

Removing and measuring full-floating wrist pins

Disassembling full-floating pins is easy and requires no special tools. Remove both snap rings and push the wrist pin free. When built-up varnish tends to stick the pin in the bore, use a drift and light hammer taps to free it.

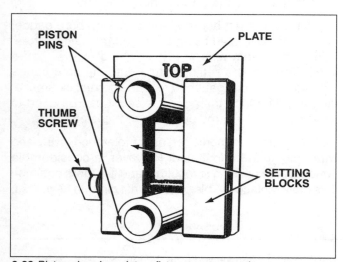

3-23 *Piston pins clamp into a fixture to measure them on a precision bore gauge.*

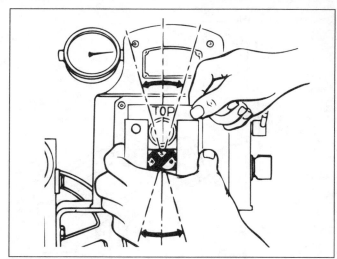

3-24 *Slip the fixture onto the bore gauge and gently rock it to get an accurate reading on the dial indicator.*

Most full-floating pins ride on a soft metal bushing that is press-fit into the small-end bore. Inspect the bushing for damage. Full-floating piston pins require a slight amount of clearance between the pin and the rod and piston bores. This clearance is generally only a few ten-thousandths of an inch. Measure the pin, piston bores, and bushing inside diameter as described for press-fit pins.

Pin bushings have a tendency to wear to a taper, figure 3-25, or develop a bell-mouthed shape, figure 3-26. If either condition exists, replace the bushings.

Installing full-floating wrist pins

Full-floating wrist pins should slip easily into the bores if the parts are properly cleaned and there is adequate

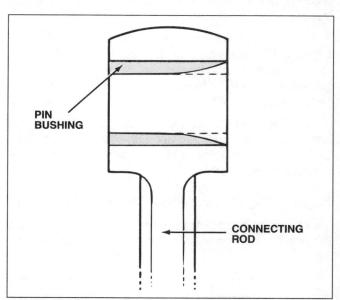

3-25 *Pin bushing taper wear results from rod misalignment and is corrected by straightening the rod and installing a new bushing.*

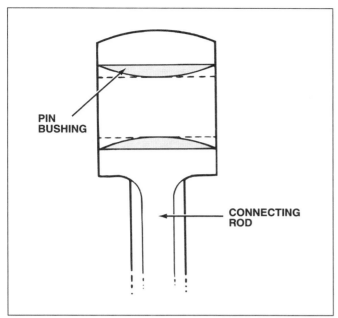

3-26 *Pin bushing bell-mouth wear is a result of too much pin clearance.*

clearance. Double check measurements to verify clearance, lubricate the parts, and slip the pin into position. Install the snap rings and make sure they seat firmly into the grooves. The open portion of the snap ring should face down, toward the crankshaft.

Connecting-Rod Reconditioning

Unless there is serious damage, connecting rods can be reconditioned and returned to service. Connecting-rod reconditioning can include straightening the rod, reducing and honing the big-end bore, and replacing bushings in the small-end bore. Check for straightness using an alignment fixture and measure the bores as described below to evaluate overall connecting-rod condition.

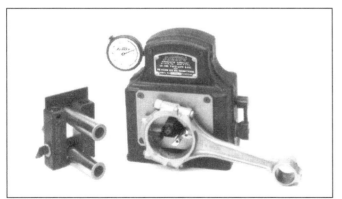

3-27 *A precision bore gauge can be used to measure connecting rod bores.*

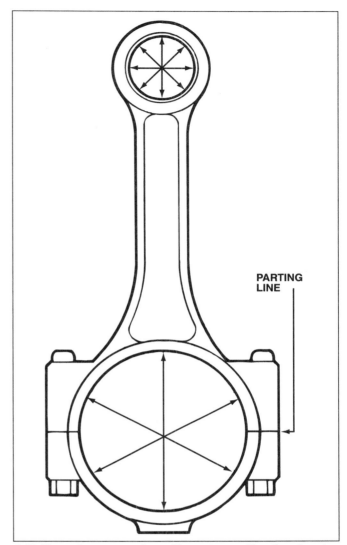

3-28 *Measuring a connecting rod bore at several locations to get accurate readings. Never measure directly across the parting lines.*

Measuring the bearing bore

The bearing, or big-end, bore of a connecting rod is subject to **stretch** and must be measured for diameter and out-of-round. Although an inside micrometer can be used, this is not a satisfactory way to check the very close tolerances specified for new connecting rods. A better way is to use either a dial-indicating device designed specifically for measuring bearing bores or a precision bore gauge, figure 3-27. Always take readings at several locations whenever you measure a bearing bore, and never measure directly across the parting lines, figure 3-28.

Two specialized machine tools, a connecting-rod grinder and a connecting-rod hone, are used to restore

Stretch: An extreme out-of-round condition on a bearing bore.

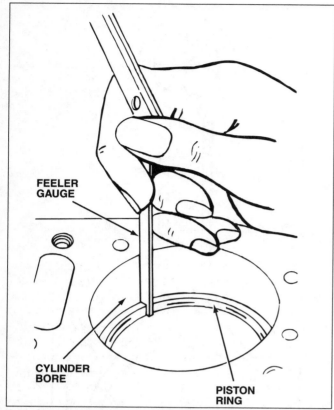

3-29 *Checking piston ring end gap with a feeler gauge.*

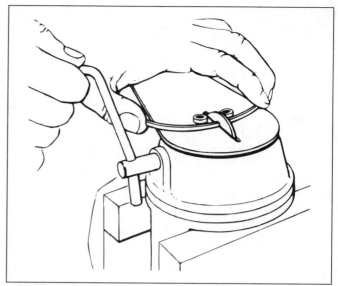

3-30 *Correcting piston ring end gap by filing on a special ring grinder.*

the bearing bore. The rod grinder removes precise amounts of metal from the parting faces of the rod and bearing cap. This reduces the inside diameter of the bore to less than standard size. Once the bore is reduced, the hone is used to open up bore and bring it back to standard size.

The connecting rod hone is also used to replace wrist-pin bushings on full-floating rods. The old bushing is pressed out, and a new bushing is pressed into the bore. An expander broach expands the bushing into the bore. Excess bushing material is trimmed away with a facing cutter. After the new bushing is installed, it is finish-honed to bring it out to final size and provide the correct amount of clearance.

PISTON RING SERVICE

The first step in installing new piston rings is to read the instructions provided with the ring set. These instructions contain key points specified by the manufacturer that often vary from the original equipment rings. Do not expect all piston ring sets to look the same or to install the same. Compression rings are often marked to indicate which side the ring faces. Although some rings are stamped "top," others use a different type of marking, such as a dot or an arrow, to indicate which way they install. As a general rule, if the

inside edge of the ring has a bevel or counterbore, it faces up, and if the counterbore is on the outside edge of the ring, it installs facing down. If only one compression ring has a bevel or counterbore on the inside, it generally goes in the top ring groove. Remember: Check instructions supplied with each set of piston rings to avoid problems.

Piston Ring End Gap

Check piston ring end gap before installing the rings on the pistons. Place the piston rings squarely in the cylinder, in the bottom of worn cylinders, or near the top, below the ridge area, of freshly reconditioned cylinders. Use a feeler gauge to check the end gap, figure 3-29. The correct end gap is approximately 0.003 to 0.004 inch for each inch of cylinder diameter, with a maximum of about 0.025 inch.

If the gap you measure is too small, file the ends of the ring to obtain the correct gap. Special ring grinders are available to open up the end gap, figure 3-30. An alternative is to use a fine-toothed file. Clamp the file in a vise and draw the ring across the file. File in one direction only, then dress any rough edges on the sides, face, and back of the ring. Recheck end gap frequently to avoid removing too much ring material.

Installing Rings on the Piston

Double check the ring side gap as described earlier, then install the rings onto the piston. Piston rings must be fitted so that the end gaps are offset from each other on the piston, figure 3-31. If the gaps are aligned vertically, serious oil consumption and blowby problems can result.

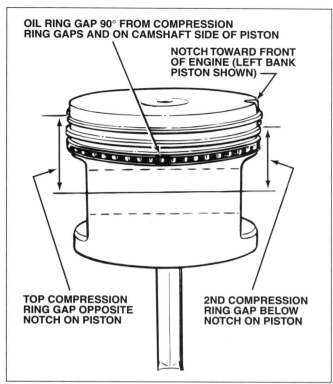

OIL RING GAP 90° FROM COMPRESSION
RING GAPS AND ON CAMSHAFT SIDE OF PISTON

NOTCH TOWARD FRONT
OF ENGINE (LEFT BANK
PISTON SHOWN)

TOP COMPRESSION
RING GAP OPPOSITE
NOTCH ON PISTON

2ND COMPRESSION
RING GAP BELOW
NOTCH ON PISTON

3-31 *Stagger piston ring end gaps to ensure good compression sealing on startup.*

Work from the bottom to the top as you install the rings on the piston. Install the oil control ring first, then fit the bottom compression ring, and install the top compression ring last. Three-piece oil control rings also install in a particular order: first, the spring expander, then, the top rail, and lastly, the bottom rail. Be sure to hold the expander securely, with the ends butted together, when fitting the rails. Never allow the ends of the expander to overlap. This will damage the cylinder wall.

Coat the ring grooves and the rings with engine oil for easier installation. Use a ring expander tool, figure 3-32, to expand compression rings so they clear the piston crown. Piston rings are brittle, and an expansion tool is used to avoid breaking them.

Installing Rods and Pistons in the Engine

Before installation, the piston rings and cylinder walls must be well oiled. Use a ring compressor tool, figure 3-33, to compress the rings so the piston will slide into the cylinder.

Remove the rod bearing cap and wipe off the bearing bore. It must be perfectly clean and dry. Place the bearing inserts into position in the rod and rod cap, making sure any lock tabs and grooves are aligned. Fit protective covers over the ends of the rod bolts to prevent accidental damage to the crankshaft journal,

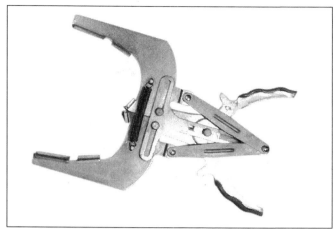

3-32 *A piston ring expander is used to install rings on the piston.*

figure 3-34. Make sure the rod and piston are both correctly assembled and positioned in the cylinder, so they both face the proper direction. A notch or arrow on the piston top will generally point toward the front of the engine. Connecting rods usually have a directional indicator, but it is often subtle, and you must know what to look for. Install the rod and piston assembly by fitting it into the cylinder and pushing down on the piston top with a hammer handle. Make sure the rod slips over the crankshaft in the normal position for assembly. Coat the bearings with a thin film of assembly lube or engine oil. Then, install the rod cap and tighten it to specified torque.

After all of the piston and rod assemblies are installed, check the connecting-rod side clearance.

Measuring connecting-rod side clearance

Side clearance is the distance between the crankshaft cheek and the side of the connecting rod, or the space

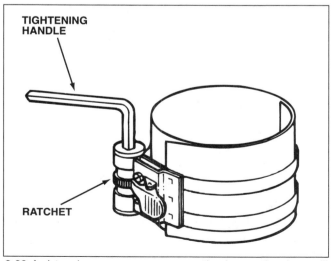

TIGHTENING
HANDLE

RATCHET

3-33 *A piston ring compressor squeezes the rings so the piston will fit into the cylinder bore.*

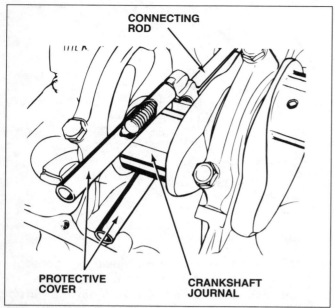

3-34 *Fit covers over the rod bolts to protect the crankshaft journal during piston installation.*

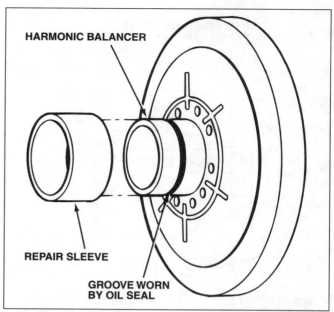

3-35 *A repair sleeve can often salvage a harmonic balancer that has a groove worn into it from an oil seal.*

between the two connecting rods that share a journal on a V-type engine. Manufacturers provide side-clearance requirements in their specification data. The largest feeler gauge that fits into the gap between the rod and crankshaft, or two adjacent rods, equals the side clearance.

HARMONIC BALANCER SERVICE

The harmonic balancer, or vibration damper is removed using a special puller. Using any other tool can damage the balancer.

Inspect the harmonic balancer for damage to the keyway and for wear to the hub caused by the timing cover seal. If the wear groove from the seal is too deep, you can usually press on a repair sleeve rather than replace the entire balancer, figure 3-35. Repair sleeves are available from gasket manufacturers. If the balancer has a rubber bonding ring, look it over carefully. These bonding rings eventually deteriorate and can cause the assembly to break apart dramatically on a running engine. Look for cracks and pieces of the ring breaking loose. Replace the balancer if any signs of damage are found.

Also inspect the crankshaft snout. Look for signs of scoring, gouging, and other damage on the shaft, keyway, and key. Dress any minor damage with a fine file, emery cloth, or stone. Replace damaged keys.

Install the harmonic balancer by aligning the key and keyway, then slipping the balancer over the snout. Use a large, open drift and hammer to seat the balancer to the crankshaft.

FLYWHEEL SERVICE

Three areas of the flywheel—the ring gear, the clutch seating surface, and the mounting flange—must be inspected to ensure good service life.

Check the ring gear for worn or missing teeth as a result of poor starter motor engagement. The ring gear is a shrink-fit on many flywheels and can easily be replaced. This is done by heating the gear, but not the body of the flywheel, with a torch flame to expand the metal. Once the gear is hot enough, it will simply drop off of the flywheel. Heat is also used to expand and install the new ring gear; a few careful raps with a hammer will guarantee a good seat.

Inspect the clutch seating surface for hard spots, heat checks, and cracks. A damaged surface can be restored by turning on a lathe or by grinding with a surface grinder. Lathe cutting will not remove hard spots. Surface grinding will, so grinding is a more common practice.

Look the mounting flange over for any nicks, chips, or burrs that will prevent the flywheel from seating onto the crankshaft. Also, check the end of the crankshaft. Correct any minor damage by dressing with a file or hand stone. Check the bolt holes for distortion. Replace the flywheel if any bolt hole distortion is found. Make sure the threaded holes on the crankshaft and flywheel are in good condition.

Flywheel Installation

The flywheel must be installed in its original position to maintain engine balance. Fit the flywheel to the

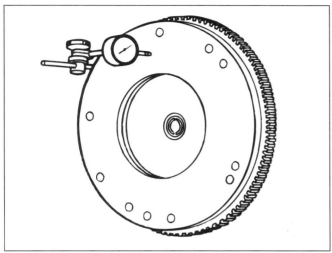

3-36 *Check flywheel runout with a dial indicator.*

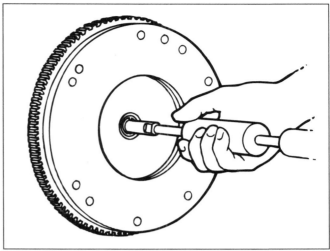

3-37 *Using a slide hammer and adapter to remove a pilot bushing.*

crankshaft flange. Install the retaining bolts, and tighten them in a star pattern to bring them to specified torque. Once it is installed, check the flywheel runout and crankshaft endplay.

Checking flywheel runout

Check flywheel runout with a dial indicator. Attach the indicator to the engine block and position its plunger so that it rests on the clutch surface of the flywheel, figure 3-36. Use a large screwdriver or prybar to pry the flywheel away from the block, then zero the indicator dial. Keep the screwdriver in place to prevent the shaft from floating, as you rotate the crankshaft one complete revolution. The highest reading on the dial indicator is total runout.

Checking crankshaft endplay

To check crankshaft endplay, attach a dial indicator to the front of the engine so that the plunger rests on the end of the crankshaft snout. Move the crankshaft as far back in the block as possible by prying with a screwdriver between the flywheel and block. Zero the dial indicator, then push the crankshaft forward in the block. The dial indicator will read total endplay.

PILOT BEARING OR BUSHING SERVICE

Pilot bearings are one of two types: a solid bushing or a sealed bearing. Bearing-type pilots are usually held in place with a snap ring. Remove the snap ring and the bearing will usually come out of the crankshaft easily. Bushings are removed using a slide hammer fitted with a blind bearing puller, figure 3-37. Bushings are installed with an interference fit and can require a considerable amount of force to remove.

Before installing a pilot bearing or bushing, inspect the crankshaft bore. The bore must be clean and free of any burrs, chips, or gouges. Check the internal bore on the bearing or bushing by sliding it onto the transmission input shaft.

All pilots, whether bearing or bushing, are installed using a special driver. For additional information on pilot bearings and bushings, refer to book three of this series.

AUXILIARY ENGINE SHAFT SERVICE

Many engines are fitted with auxiliary shafts. These include balance, intermediate, idler, counterbalance, or silencer shafts. Auxiliary shafts must be removed and inspected during engine rebuilding. In an inline engine, auxiliary shafts are generally driven by the crankshaft through a timing chain or belt. In V-type engines, auxiliary shafts are gear driven by the camshaft. Be sure timing marks align before disassembling.

Remove the shafts, then inspect and measure the bearing journals. Check for diameter, taper, and out-of-round as previously described. Check shaft straightness using V-blocks and a dial indicator. Auxiliary shafts may use split shell bearings or press-fit insert bearings. Inspect, measure, and replace bearings, as necessary, using techniques described earlier in this book.

To reinstall the shaft, be sure all timing marks align, and fit the parts following the procedures provided by the engine manufacturer. Rotate the crankshaft two complete revolutions, and recheck to make sure all of the timing marks align.

1. Connecting rods are marked:
   a. After removal
   b. With a scribe
   c. Before removal
   d. On the bottom of the cap

2. After cleaning the block in a hot tank, you should:
   a. Remove the ring ridge
   b. Brush clean the oil galleries
   c. Remove the core plugs
   d. Mark the bearing caps

3. Cylinders are measured near the top and near the bottom to determine
   a. Diameter
   b. Out-of-round
   c. Taper
   d. Oversize

4. A typical repair for a cast-iron block with 0.024 inch of cylinder taper would be:
   a. Bore to 0.025 over and install new pistons
   b. Deglaze the cylinders and install new rings
   c. Hone, then ream to 0.030 over and install new pistons
   d. Bore, then hone to 0.030 over and install new pistons

5. After machining, the cylinder walls are cleaned with:
   a. Solvent
   b. Soap and water
   c. A hot tank
   d. A glaze breaker

6. Crankshaft journals are inspected for:
   a. Taper
   b. Twist
   c. Parallelism
   d. All of the above

7. Which of the following is **NOT** a typical crankshaft journal wear pattern?
   a. End-to-end taper
   b. Hourglass taper
   c. Fillet taper
   d. Barrel taper

8. If the block and rods are in good shape, main and rod bearing insert wear will be:
   a. Even across the bearing face
   b. Concentrated on apparently random areas of the bearing faces
   c. In a short arc near the fillet
   d. Greatest in the center and minimal at the parting line

9. Which of the following does **NOT** indicate a bent connecting rod?
   a. Uneven side clearance on the piston skirt
   b. Wear along one edge of the bearing shells
   c. Wear at opposite ends and sides of the bearing shells
   d. A diagonal pattern on the thrust face of the piston

10. A connecting-rod alignment fixture is used to check all of the following **EXCEPT**:
    a. Offset
    b. Stretch
    c. Bend
    d. Twist

11. Press-fit piston pins are installed with about:
    a. 0.003 inch of interference
    b. 0.002 inch of clearance
    c. 0.001 inch of clearance
    d. 0.001 inch of interference

12. Standard procedure to repair connecting rod big-end bore stretch is to:
    a. Bore to oversize
    b. Hone to oversize
    c. Machine back to standard size
    d. Fit undersize bearings

13. Typical piston ring end gap for a 3.985-inch diameter piston would be:
    a. 0.003 to 0.004 inch
    b. 0.006 to 0.008 inch
    c. 0.009 to 0.012 inch
    d. 0.012 to 0.016 inch

14. When assembling the connecting rod onto a piston:
    a. The rod can be installed in either direction
    b. The rod can be installed in one direction only
    c. The piston pin must be installed in the connecting rod first
    d. The piston notch is always to the rear of the engine

15. Install rods and pistons:
    a. By tapping with a ball peen hammer
    b. By pushing gently with a wooden hammer handle
    c. Clean and dry
    d. With rod caps reversed

Chapter Four

LUBRICATION AND COOLING SYSTEM DIAGNOSIS AND REPAIR

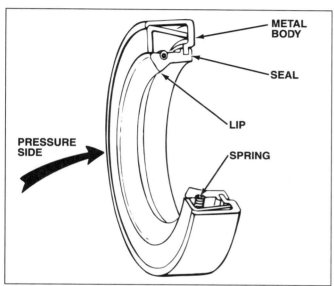

4-1 *A one-piece lip seal assembly makes an excellent shaft seal because oil pressure pushes the sealing lip into the shaft.*

This chapter details the various inspections, tests, and repair procedures for the lubrication and cooling systems of the engine.

LUBRICATION SYSTEM

Besides changing the oil and oil filter at regular intervals, lubrication system service generally consists of:

- Locating and repairing leaks
- Testing the oil pressure
- Servicing the oil pump.

Gaskets and Seals

Cork, rubber, or composite material gaskets are used where surfaces may not be perfectly flat and where considerable crush is required. Oil pans, valve covers, and timing covers generally use cork, rubber, or composite gaskets. Cork gaskets change shape with the humidity in the air, and they may not fit properly when first removed from the package. They can be made to grow by soaking them in warm water, and made to shrink by drying them over a warm surface. Gasket

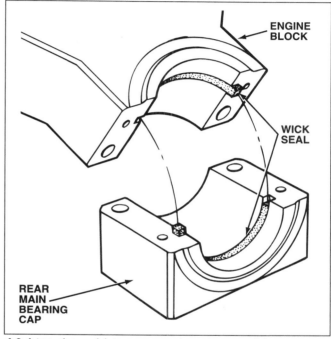

4-2 *A two-piece, wick-type, rear, main seal presses into grooves on the bearing cap and engine block.*

sealer may be used on the sheet metal side of the gasket to hold it in place during installation. Flatten the sheet metal sealing surfaces with a ball peen hammer before you install the gaskets.

Lip-type oil seals, figure 4-1, are used where movement occurs against the seal surface. These seals have a definite lip which should face the lubricant. A spring garter is generally used behind the lip to maintain pressure and improve sealing. Coat the outer edge of the seal with gasket sealer to prevent oil leaks around the outside, and lubricate the seal lip with engine oil before installing it.

Wick-type oil seals are impregnated with a graphite lubricant. An example is the rear, main oil seal. The seal is installed in a groove in the block and in a groove in the bearing cap or bearing seal, figure 4-2. The seal is seated in place, and the excess length is trimmed flush with the parting line.

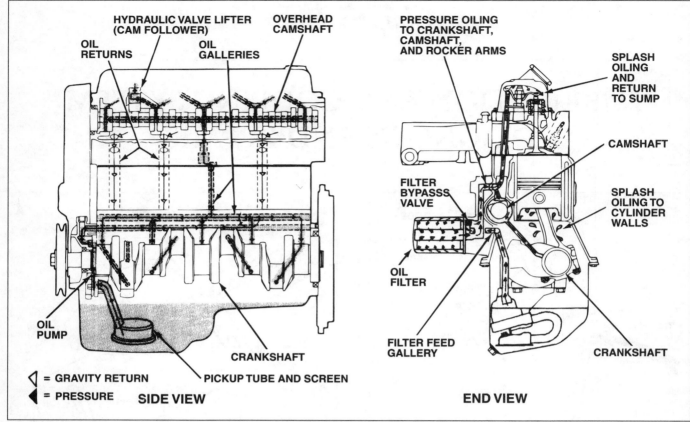

4-3 *Excessive leakage or clearance anywhere in an engine oiling system reduces oil pressure and causes engine damage.*

Gasket sealants

Many cork and rubber gaskets can be installed without sealants, and a thin coating of sealant is all that is required to seal a paper gasket. There are a wide variety of products available to do the job. Generally, only non-hardening gasket sealants should be used in automotive repair. Since these sealants remain pliable, the seal is not compromised by different rates of metal expansion during operation.

A gasket may not conform to all the channels and ridges of the flange. Using a formed-in-place sealant will fill these irregularities. There are two types of this sealant—aerobic and anaerobic.

Aerobic, or room-temperature vulcanizing (RTV), sealant cannot be used on the exhaust system because it cannot withstand such high temperatures. You must fit the parts together within 10 minutes of applying the sealant because it will harden.

Anaerobic sealants cure only after the mating parts are bolted together, which excludes air from the joint.

This sealant is thinner than RTV compound and is not practical for flexible covers. It is recommended for machined surfaces only.

You must be aware that the electronic controls on some engines can be affected by certain gasket sealing compounds. Sealants cure in a running engine and, as they do, spent chemicals are emitted. These contaminants can circulate in the engine and cause faulty signals from various sensors. Oxygen sensors are especially vulnerable to certain RTV silicone sealers. Use the correct type of sealant, and use it only where specified.

Oil Pressure Testing

The following tests or diagnostic inspections may be performed on assembled engines in the vehicle to locate the cause of low oil pressure or noises. Excessive clearance or leakage anywhere in the lubrication system can cause a loss of oil pressure, figure 4-3.

Aerobic: Curing in the presence of oxygen.

Anaerobic: Curing in the absence of oxygen.

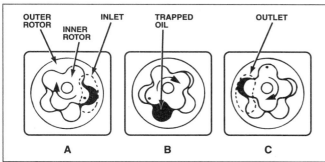

4-4 *A rotor pump uses lobes to pick up oil from the inlet port (A), trap it between the two rotors (B), then compress and push through the outlet under pressure (C).*

Checking oil pressure

Engines that indicate low oil pressure on the instrument cluster can be tested more precisely by connecting a mechanical oil pressure gauge. With the engine running, the gauge will display the true oil pressure the engine is producing.

Another test requires removing the oil pan, connecting an outside source of oil pressure, and turn the engine slowly by hand. While the engine is turning, carefully watch the oil flow through all pressure-fed bearings, including the camshaft bearings. If excessive flow is found, rotate the crankshaft 45 to 90 degrees in either direction and recheck. A false indication of excessive flow can occur if the crankshaft and engine block **oil galleries** and the oil holes in the bearings are aligned to allow unrestricted flow. If rotating the engine causes the flow to decrease, this alignment is probably the cause of the excessive flow. Also, check the oil pump pressure relief valve to see if it is stuck whether the open position.

Oil Pump Service

There are two basic oil pump designs: rotor type, figure 4-4, and gear type, figure 4-5. Gear-type pumps generally bolt to the engine block or to a main bearing cap. These pumps are driven by a shaft that is geared to the camshaft or to a layshaft. Some rotor-type pumps are similar in design and take their power off the camshaft. Others mount to the timing cover and are driven directly by the crankshaft.

An oil pump must be removed from the engine and cleaned before it can be inspected. In most cases, the oil pan must be removed to access the oil pump. With

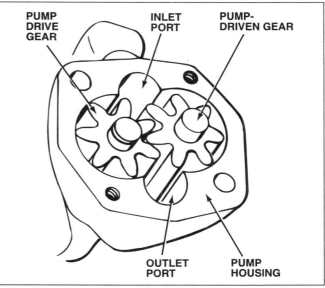

4-5 *A gear pump uses the mesh of two straight-cut gears to pick up, pressurize, and deliver oil to the engine.*

a crankshaft-driven pump, the timing cover must be removed to access the pump.

Once the oil pump is removed from the engine, disassemble it, thoroughly clean the parts with solvent, dry with compressed air, then inspect the components for wear and damage. It is especially important that oil pressure relief valves and oil passages be free of **sludge** and **varnish**, so the valve can move freely in its bore and operate normally. Also, thoroughly clean and carefully inspect the pickup screen for damage before reinstalling it.

If the pump is shaft-driven, begin your inspection by examining the driveshaft. Look for signs of twisting, bending, or other distortion. Replace the shaft if anything appears questionable. The end of the shaft must fit snugly into the pump with virtually no free play. If the end is worn or rounded off, install a new driveshaft.

All oil pumps use a pickup tube and screen to filter large particles out of the intake oil. Any debris that can pass through the screen is small enough to go through the pump without locking up the gears or rotors. The pickup may be bolted, threaded, or pressed onto the pump body or engine block. Remove the pickup assembly, wash it in solvent, and inspect the screen. If the screen shows any sign of damage, replace it.

Oil Galleries: Pipes or drilled passages in the engine block that are used to carry engine oil from one area to another.

Sludge: Black, moist deposits that form in the interior of the engine. A mixture of dust, oil, and water whipped together by the moving parts.

Varnish: A hard, undesirable deposit formed by oxidation of fuel and motor oil.

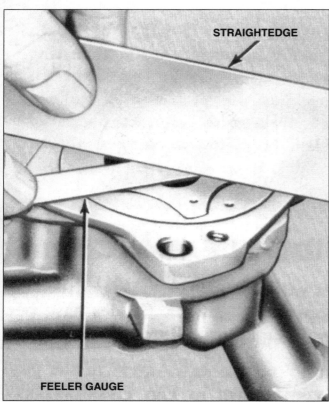

4-6 *Checking oil pump end clearance with a straightedge and feeler gauge.*

Gear-type oil pumps

Gear-type pumps use the meshing of two gears, a drive gear and an idler gear, to provide pressurized oil flow to the engine. The gears fit into the pump housing and are held in place by the end cover. Remove the end cover to access the gears. Make sure there are clear index marks on the gears before you remove them from the pump body.

Inspection and measurement

There are two measurements used to determine whether a gear pump is in serviceable condition: end clearance and gear-to-housing clearance. End clearance is the distance from the gear faces to the end cover and is measured with a feeler gauge and a straightedge. Place the pump on the workbench with the gears facing up and lay the straightedge across the pump. Clearance is equal to the largest feeler gauge that will fit between the gear faces and the straightedge, figure 4-6. In general, end clearance should be less than 0.003 inch.

Gear-to-housing clearance equals the largest feeler gauge you can fit between a gear tooth and the housing. Measure both gears at several locations and rotate the gears to get readings on different teeth. Typically, gear-to-housing clearance should not exceed 0.005 inch.

Remove the gears and look over all the contact surfaces of the gears, pump body, and end cover for signs of wear. Replace the pump if you find any deep scoring, chips, cracks, or other deformation. If the pump body is in good condition, you can install a rebuild kit and return the unit to service. A rebuild kit includes new gears, pressure relief valve spring, seals, and gaskets.

The end plate of the pump housing develops a wear pattern and slight scoring from contacting the rotors under normal operating conditions. You can resurface the end plate if scoring is less than 0.003 inch deep. Resurface the plate by lapping on a flat surface using wet or dry sandpaper lubricated with engine oil.

Rotor-type oil pumps

All rotor pumps work on the same principle, but there are a number of different rotor designs in use. With a rotor pump, an inner rotor is positioned off-center in the pump body and driven by an external power source, such as a driveshaft or the crankshaft. The outer rotor fits around and is driven by the inner rotor. The clearance between the two rotors is constantly changing. Oil is carried from large to small clearance areas between the rotors, then forced from the pump outlet under pressure.

Most rotor pumps can be readily disassembled for inspection. If the pump body is in good condition, install a rebuild kit and return the unit to service. A rebuild kit includes new rotors, pressure relief valve spring, seals, and gaskets.

Remove the end cover to access the rotors. Check for index marks on the rotor faces. Although the rotors do not have to be aligned with each other on assembly, they both must face their original direction.

Inspection and measurement

Three measurements are required on a rotor pump: end clearance, rotor-to-rotor clearance, and rotor-to-housing clearance. All measurements are taken with a feeler gauge. End clearance is the distance from the rotor faces to the end cover and is measured using a straightedge as described for gear-type pumps.

To check rotor-to-rotor clearance, position the rotors so that a lobe on the inner rotor faces a lobe on the outer rotor, then measure between the lobes with a feeler gauge, figure 4-7. As a general rule, rotor-to-rotor clearance should be less than 0.010 inch.

Remove the inner rotor to measure rotor-to-housing clearance. Then, insert feeler gauge blades between the outer rotor and the pump body to measure the clearance, figure 4-8. Rotor-to-housing clearance equals the thickest feeler gauge blade that will fit the

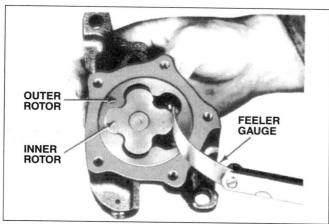

4-7 *Align inner and outer lobe tips and check oil pump rotor-to-rotor clearance with a feeler gauge.*

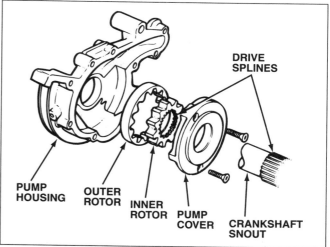

4-9 *The oil pump inner rotor splines to the crankshaft, and the pump assembles behind the timing cover on some engines.*

gap. If clearance exceeds 0.012 inch, replace the pump or install a rebuild kit.

Check all parts for scoring, chips, cracks, or other deformation and install a rebuild kit or replace the pump if any is found. You can resurface the end plate as previously described for gear pumps.

Crankshaft-driven oil pumps

Crankshaft-driven pumps are rotor-type pumps that mount to the engine timing cover. The inner rotor slips over the crankshaft snout and is secured to the shaft by splines. The outer rotor fits into a pocket on the inside of the timing cover, and the pump assembly is held in place by a cover plate that bolts to the timing cover, figure 4-9. The service procedures for crankshaft-driven pumps are similar to those outlined for rotor pumps. However, tolerances are generally tighter.

Inspection and measurement

It is often impossible to measure end clearance using a straightedge because the pump assembly is

recessed into the front cover. Check end clearance, also known as rotor drop, using a depth micrometer. Position the micrometer so that it rests on the cover mounting flange and straddles the opening, figure 4-10. Run the micrometer spindle down until the anvil contacts the rotor face, then take a reading. Acceptable tolerance is generally in the 0.001- to 0.004-inch range.

Measure rotor-to-rotor and rotor-to-housing clearances as previously described. Expect rotor-to-rotor clearance, also known as tip clearance, readings in the 0.004- to 0.009-inch range. Rotor-to-housing, or side clearance, should be slightly more, with an upper limit of about 0.015 inch.

Oil pressure relief devices

The pressure relief valve is a component that is common to all oil pumps. Usually, a spring-loaded plunger

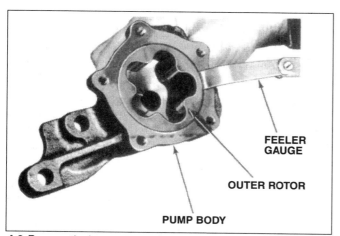

4-8 *Remove the inner rotor to check oil pump rotor-to-housing clearance with a feeler gauge.*

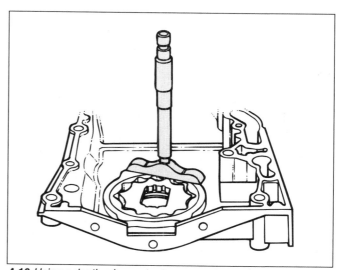

4-10 *Using a depth micrometer to measure rotor drop on a crankshaft-driven oil pump.*

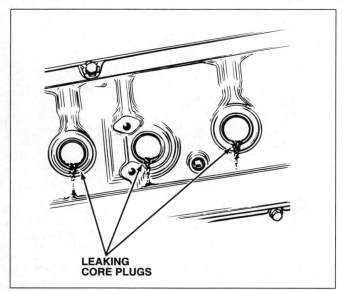

4-11 *Check for signs of coolant leakage at the core plugs on the sides of the engine block.*

bleeds off excess oil to maintain optimum pressure to the engine. The relief valve, which may be installed in the pump housing, timing cover, or oil filter flange, is held in place by a threaded plug, cotter pin, roll pin, snapring, cover plate, or other fastener.

Remove the fastener that holds the relief valve in place and withdraw the plunger and spring from the bore. Pay close attention to how the pieces are assembled; installing them incorrectly can result in no or excessive oil pressure. Wash the parts in clean solvent and inspect for signs of scoring or excessive wear. Clean the bore with a suitable brush. Measure the spring length and tension, compare to specifications, and replace as necessary.

Oil pump installation

When assembling an oil pump, remember that the gasket that seals the end housing to the pump body is manufactured to a specific thickness to provide the proper end clearance. Do not substitute any other gasket. Too much end clearance results in an oil pressure loss, while too little clearance can cause binding and premature failure. Install the gasket without sealant. Using sealant can alter clearance. In addition, excess sealer can be drawn into the pump, restrict oil flow, and lead to engine damage.

You must prime the oil pump before you install it. If not primed, the pump will not circulate oil immediately on startup, or it may not be able to pick up oil at all.

Submerge the pump pickup in a container of clean engine oil and spin the rotors or gears by hand until the pump discharges a good stream of oil. Prime the pump with engine oil only. Do not pack a pump with grease, assembly lube, or other heavy lubricant unless specifically instructed to do so by the engine manufacturer.

After the pump is installed, the pickup screen must be positioned parallel to the bottom of the oil pan, with about ¼ to ⅜ inch of clearance between the screen and the pan. Measure clearance by taking two readings with a ruler. Measure from the screen opening to the block pan rails, then from the floor of the sump to the flange on the oil pan. The difference between the two measurements is the clearance.

COOLING SYSTEM SERVICE

Basic cooling system service consists of:

- System inspection
- System and component testing
- System cleaning
- Component replacement.

Cooling System Inspection

To maintain normal operating temperatures, all of the cooling system components must be in good condition. Listening while the engine is running will often reveal cooling system problems. Listen for:

- An engine thump at normal temperature caused by a restriction in the **water jacket** or an incorrectly installed head gasket.
- A screeching noise caused buy a loose drive belt.
- A buzz or whistle caused by a poor pressure cap seal or vibrating radiator fan shroud.
- A ringing or grinding noise from a worn or damaged water pump bearing or loose drive belt pulley.
- A gurgling from the radiator caused by a plugged radiator or air in the coolant.

With the engine off, evaluate system condition by performing the following checks:

- The water pump drive belt must be correctly tensioned and free of any glazing, deterioration, or other damage.
- Look at all hoses for hardness, cracks, or brittleness, softness or interior damage, and loose connections or leakage.
- Look for signs of leakage around the core plugs, figure 4-11.

Water Jacket: The area in the block and head around the cylinders, valves, and spark plugs that is left hollow so the coolant can circulate.

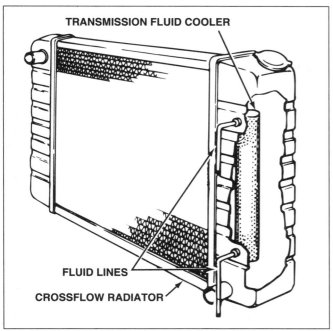

TRANSMISSION FLUID COOLER

FLUID LINES

CROSSFLOW RADIATOR

4-12 *Check for automatic transmission fluid cooler leakage and signs of damage to the lines and fittings while inspecting the radiator.*

- Inspect the radiator for oil, rust, or scale inside the filler neck, leakage at the tank seams, kinked or damaged overflow tubes, and damage around the automatic transmission cooler lines, figure 4-12.
- Inspect the radiator cap; it must fit securely on the filler neck. Also, check for a brittle cap seal, figure 4-13, insufficient spring action, and correct pressure rating.
- Inspect the heater core for signs of leakage and loose connections.
- Look at the water pump for leakage at the shaft seal, bleed hole, and gasket. Check the pump for wobble, binding, or looseness when turned by hand.
- Look for cooling fan blade misalignment or damage.
- Check fan clutches for fluid leakage, noise or roughness when turned by hand, and excessive shaft endplay.
- Check electric cooling fans for loose electrical connections and damaged wiring.

Cooling System Testing

Testing of the cooling system generally consists of testing the coolant and performing system and radiator cap pressure tests. You may also want to test the operation of the thermostat.

Testing the coolant

Coolant concentration and effectiveness is tested with

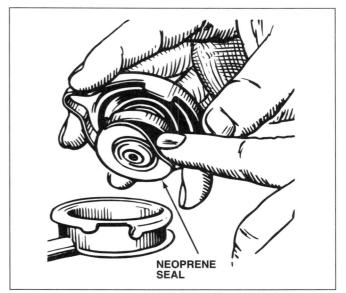

NEOPRENE SEAL

4-13 *Make sure the seal on the radiator cap is soft and pliable.*

a cooling system **hydrometer**. For accurate results, the coolant should be hot when tested. Before doing the test, draw a coolant sample into the hydrometer and return it to the radiator a few times to stabilize the internal thermometer of the hydrometer. Test as follows:

1. Hold the hydrometer straight and draw enough coolant to raise the float, figure 4-14. The float should not touch the sides of the hydrometer.
2. Take the reading at eye level, watching the top of the letter on the float that is touched by the coolant.
3. Find this letter on the hydrometer scale and read down the column under the letter until you are opposite the thermometer reading.
4. The number shown at this point is the degree of protection given by the coolant in the system.

Electrolyte testing

As coolant deteriorates, it turns into a weak electrolyte which can corrode aluminum surfaces. To determine whether your coolant has turned into an electrolyte, insert the positive lead of a voltmeter into the fluid. Attach the negative meter lead to the neck of the radiator. The voltage reading should not exceed 0.3 volt. If it does, drain and flush the cooling system and fill with fresh coolant.

Pressure testing

A pressure test is performed on a cold engine that is not running. Remove the pressure cap to test both the cooling system and the cap. Test procedures are detailed in Chapter One of this book.

Hydrometer: A device used to measure the specific gravity of a fluid.

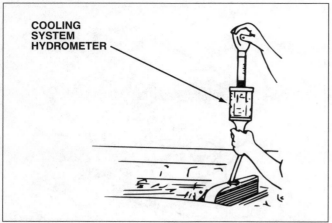

4-14 *A hydrometer checks the concentration and effectiveness of the coolant in the system.*

On-car thermostat test

Thermostat opening temperature can be checked without removing the thermostat. Perform the test on a cold engine. Remove the radiator pressure cap and check and correct the coolant level. Then, test the thermostat as follows:

1. Place a thermometer, or a temperature-registering label, in the filler neck of the radiator.
2. Start the engine and monitor the thermometer as the engine warms up.
3. Watch the coolant that is visible at the filler neck. As the thermostat opens and you see the coolant swirling around, note the thermometer reading.

If the thermostat opened above or below its temperature rating, the unit is defective and should be replaced.

Electric fan test

An electric cooling fan that fails to engage is often the result of a defective coolant temperature switch. Check the switch as follows:

1. With the engine cold, disconnect the electrical connector from the coolant temperature switch. Connect an ohmmeter, or a self-powered test lamp. Most switch circuits are open when cold.
2. Reconnect the switch connector. Start the engine and run it to normal operating temperature.
3. If the fan does not engage, switch off the engine. Disconnect the switch connector and retest the circuit. If an open circuit is found, replace the switch.

The fan motor can be checked quickly using a jump wire. Disconnect the electrical connector from the coolant temperature switch and connect the jump wire across the terminals of the connector. Switch on the ignition, and the fan should run.

Clutch fan test

The operation of a fan clutch can be checked using a timing light and tachometer. Perform this procedure on a cold engine. Attach a thermometer to the engine side of the radiator, so you can watch it during testing to prevent overheating. Perform the test as follows:

1. Connect the timing light and tachometer, then start the engine.
2. Aim the timing light at the fan blades; they should appear to move slowly.
3. Block the radiator to restrict air flow and promote rapid heat buildup. Keep an eye on the thermometer and do not allow the engine to overheat.
4. When the thermometer indicates the fan clutch engagement point, remove the radiator cover and aim the timing light at the fan blades. Fan speed should increase as the clutch engages, and the blades will appear to move faster in the timing light beam.
5. Continue to watch the fan blades with the timing light as the coolant temperature drops. Fan speed should decrease once the temperature falls below the engagement point.

Replace the fan clutch if it fails to engage or if it engages and disengages at an incorrect temperature.

Cooling System Cleaning

The cooling system should be completely drained and flushed once a year to prevent internal damage. If the system is regularly serviced and only light contamination is present, the system can be flushed with clean water.

1. Drain the old coolant either by opening the radiator and engine drain plugs, or by disconnecting the lower radiator hose.
2. Remove the thermostat and reinstall the thermostat housing.
3. Place a hose in the radiator filler neck and adjust the water flow to keep the water level at the top of the radiator while water is flowing out of the drains.
4. Flush the system for 10 minutes. Run the engine at idle for a more thorough flushing.
5. Reinstall the thermostat and close all the drains or reconnect the lower radiator hose.
6. Fill the system with the recommended amount of coolant to the correct level.

If the system is badly contaminated, it will need to be flushed with a chemical cleaner. The system is drained and refilled with clean water, then the chemical cleaner is added. Run the engine at fast idle for about 30 minutes. Do not allow the coolant to boil. Stop the engine and drain the system while the engine is still

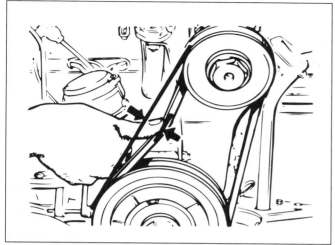

4-15 *Checking drive belt tension by deflection.*

warm. Allow the engine to cool. Close the drains and fill the system with the recommended amount of neutralizer and water. Run the engine at fast idle for about 10 minutes, then stop the engine and drain the system. Reinstall the thermostat and fill the system.

Many late-model vehicles require bleeding to remove all of the trapped air when the cooling system is drained and refilled. These procedures vary by manufacturer, model, and engine. Check the appropriate service manual. Failure to properly bleed a cooling system can result in overheating.

Component Replacement

Cooling system components that periodically need replacement include:

- Drive belts
- Hoses and thermostat
- Core plugs
- Water pump
- Radiator.

Replacing drive belts

Never force or pry a belt over pulley flanges. If the belt cannot be run into the grooves by rotating the pulley, move the driven accessory to obtain closer centers. With the belt removed, examine the pulleys for damage and misalignment. Replace pulleys as required and install the new belt.

Belts must be properly tensioned. A loose belt will slip, and a tight belt can damage bearings. Adjust tension to specifications by either the deflection method, figure 4-15, or using a strand tension gauge, figure 4-16. Tighten retaining bolts securely to maintain proper adjustment. Start the engine and watch how the belt rides in the pulley grooves. Ideally, the belt should be flush with or not more than $1/16$ inch above the top of the

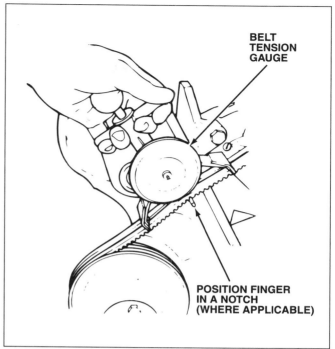

4-16 *Checking drive belt tension with a gauge.*

pulley grooves. If it is too high, the belt sides will wear excessively; if it is too low, the belt will bottom on the grooves, wear prematurely, glaze, and eventually slip. Make sure the belt cross-section conforms to the angle of the pulley grooves. If it does not, recheck to be sure the correct belt was installed.

Visually inspect pulleys for signs of damage and rotate them with the belt removed to check for free movement of the bearings. Replace the pulley, or bearings, if any damage is found or if there is binding on rotation. Also, check for proper alignment between the pulleys. Pulley alignment is especially critical for serpentine belts. Incorrect pulley alignment loads the sides of a belt and leads to premature failure.

On engines that have an automatic belt tensioning device, inspect the tensioner whenever the belt is removed. Make sure the bearing is in good condition and allows the tensioner to rotate smoothly. Look for nicks, dings, or scratches that might damage the belt. With the belt installed, check belt tension to make sure the spring assembly is applying enough pressure to keep the belt taut.

After installing a new drive belt, allow the engine to run for at least 10 to 15 minutes. Switch the engine off, then recheck and adjust belt tension using "used" belt specifications.

Radiator and heater hose service

Replace any hose that is brittle, cracked, or swollen. A hose that has defects on its outer surface will probably

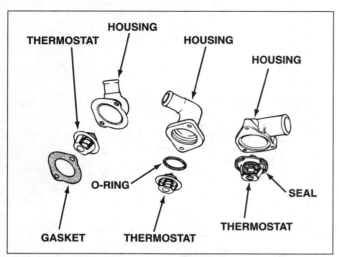

4-17 *Always replace the gasket or seal with a new one when installing a thermostat.*

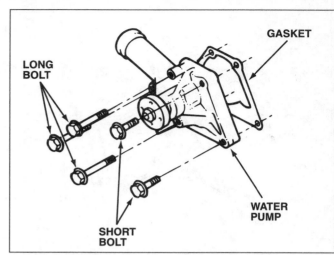

4-18 *Water pumps often attach to the block with different length bolts. Be sure to install them correctly.*

be defective inside as well. Be sure the replacement hose is the same configuration as the old one. Trim the new hose to the proper length as needed. The installed hose must be free of kinks and twists. Draw the hose clamps up tight, but do not overtighten them to the point where they cut into the hose.

Thermostat service

To access the thermostat, disconnect the radiator hose, remove the thermostat housing, and lift out the thermostat. Thermostats seal with either a gasket, O-ring, or rubber seal, figure 4-17. The seal, no matter which type is used, is always replaced when the thermostat is removed. Scrape off all traces of old gasket and sealant from both sealing surfaces. Once removed, the thermostat can be bench-tested as follows:

1. Suspend the thermostat and a thermometer in a heatproof container and fill the container with water.
2. Heat the water. Note the temperature when the thermostat begins to open and the temperature when it is fully open.
3. Turn off the heat source. Note the temperature at which the thermostat is fully closed.

The thermostat should begin to open when the temperature is about 3 to 9 degrees below the rated temperature. The thermostat should be fully open when coolant temperature matches the rating. As it cools down, the thermostat should be fully closed when the coolant reaches the same temperature where opening began. Replace the thermostat if it did not perform as described.

Removing and replacing core plugs

Any signs of moisture, rust, or coolant stains around a core plug indicates seepage. The plug should be replaced. A pressure test of the cooling system is a very

effective way to test for leakage. If pressure bleeds down, but you cannot spot a leak, feel for dampness around the core plugs.

Core plugs can be removed using a punch and hammer. Place the end of the punch on the edge of the plug and strike the punch with the hammer to turn the plug sideways in the bore. Be careful not to gouge or damage the sides of the plug bore with the punch. Once the plug is sideways, you can pry it out with a screwdriver or pull it free with a pair of pliers. Clean the sides of the bore with emery paper.

Cup-type plugs have two different shoulder heights: deep cup and shallow cup. The two designs cannot be interchanged. To select the correct depth, measure the thickness of the core holes.

Before installation, coat the sides of the core plug with gasket sealer. The plug is driven into the bore using a special installation tool. The cup must fit squarely in the bore. When properly installed, the edges of the cup will be slightly recessed into the engine block.

Water Pump Service

Here are some general tips to remember when replacing a water pump:

- First, drain the coolant.
- Note the locations of bolts of different lengths. These must be returned to their original position, figure 4-18.
- Use thread sealant when reinstalling bolts.
- Make sure the pump gasket is aligned so that coolant bypass holes are not blocked.
- Be sure all mating surfaces are perfectly clean and use only non-hardening sealants.

- Compare the old pump to the replacement before installing it. Be sure the pulley is correctly positioned to maintain belt alignment.

Water pump replacement often requires a considerable amount of disassembly to access the pump. Be sure that all the parts that were removed are reinstalled and properly secured.

Radiator removal and replacement

Radiators generally need repairs or replacement only after years of service or neglect. Infrequent cooling system flushing can cause built-up sediment and contaminants to block the internal passages of the radiator and restrict coolant flow. A clogged radiator must be removed and disassembled to be repaired. Normally, **rodding** the passages is sufficient to return the unit to service. In extreme cases, the radiator core will need to be replaced. Both services are generally performed at a specialty shop.

When installing radiators, be sure that all rubber mounts are in good condition. Transmission cooler tubes and hoses must be securely attached with all the mounting brackets in place. Also, check all hose and transmission cooler connections for tightness. Check for leakage after running the engine.

Rodding: A method of repairing a radiator by running a drill bit through the internal passages of the core.

1. Of the following types of sealant, which would be used when replacing a valve cover with a formed-in-place gasket?
   a. High tack
   b. Anaerobic
   c. Aerobic
   d. Metabolic

2. A false indication of excessive oil flow can occur if:
   a. The oil galleries in the block and the oil holes in the bearings are aligned
   b. There is excessive valve lash
   c. The warning lamps or indicators are malfunctioning
   d. Oil dilution has occurred from a rupture in the fuel pump diaphragm

3. When installing an oil pump, it should be primed with:
   a. Assembly lube
   b. Petroleum jelly
   c. Motor oil
   d. High-pressure lubricant

4. Oil pump tip clearance is the same as:
   a. Rotor drop clearance
   b. End clearance
   c. Rotor-to-housing clearance
   d. Rotor-to-rotor clearance

5. An oil pressure relief may be located in any of the following locations **EXCEPT**:
   a. Oil pump housing
   b. Oil filler neck
   c. Oil filter flange
   d. Timing cover

6. With the engine running, an engine thump at normal temperature can mean:
   a. A loose drive belt
   b. A poor pressure cap seal
   c. A restriction in the water jacket
   d. A worn water pump

7. A V-type drive belt that rides too low and bottoms in the pulley groove causes:
   a. Premature wear, glazing, and slippage
   b. The belt sides to wear excessively and split
   c. High belt tension that can lead to bearing damage
   d. The belt to bind, seize, and eventually break

8. The rated temperature of a thermostat indicates:
   a. The temperature where it begins to open
   b. The temperature where it is fully open
   c. The temperature where it fully closes
   d. The thermostat bypass temperature

9. A thermometer, timing light, and tachometer can be used to check the operation of:
   a. The cooling system
   b. The thermostat
   c. An electric fan
   d. A fan clutch

10. A pressure tester can be used to:
    a. Test thermostats
    b. Test radiators, pressure caps, and hoses
    c. Test for exhaust leaks into the cooling system
    d. Test the coolant

Chapter Five

FUEL, ELECTRICAL, IGNITION, AND EXHAUST SYSTEMS INSPECTION AND SERVICE

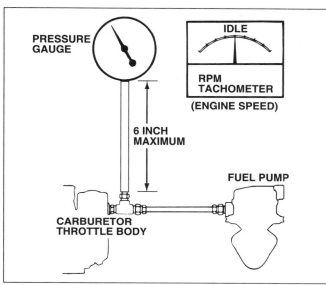

5-1 *Pressure gauge connection for pressure testing a mechanical fuel pump.*

This chapter discusses various components of the subsystems that are required on a running engine. Many of these systems are addressed in other ASE certification examinations, and successfully passing the A1 test requires only general or basic knowledge of these systems. Therefore, the following discussions are brief and do not cover all aspects of each system. For additional information, refer to the appropriate volume in this series.

FUEL SYSTEM TESTING AND SERVICE

A constant flow of clean, pressurized fuel and filtered air must be supplied to ensure proper combustion. These requirements are handled by the fuel and air supply systems. A fuel pump must deliver enough fuel at the correct pressure to the engine in order to maintain efficient combustion. Two types of fuel pumps are used:

- mechanical
- electric.

Both types can usually be checked for pressure and volume.

Mechanical Pump Testing

Mechanical fuel pumps, which are used on some carbureted engines, are **positive-displacement pumps**. A mechanical fuel pump is checked for delivery pressure and volume.

Pressure test

Pressure testing is performed using a vacuum-pressure gauge. For an accurate gauge reading, hold the vacuum-pressure gauge at the same height as the carburetor. Connect the gauge to the carburetor using a 6-inch, or less, length of fuel hose. If there is a fuel return line on the pump, clamp it closed. To test:

1. Remove the air cleaner and connect a tachometer to the engine.
2. Disconnect the fuel inlet line at the carburetor and attach the vacuum-gauge line. The gauge can also be connected with a tee fitting, figure 5-1.
3. Start and run the engine at the specified rpm. Compare the pressure reading to specifications.
4. Stop the engine. Pressure should hold for several minutes.

Volume test

To check fuel pump volume:

1. Disconnect the fuel supply line at the carburetor and attach a length of fuel hose, figure 5-2. If a fuel filter is used, do not remove it.
2. Place the free end of the tubing into a graduated container.
3. Start the engine and run it at 500 rpm for 15 seconds while monitoring fuel level in container.
4. Note the fluid level at 15 seconds, then multiply by four to get a one-minute reading.

Positive-displacement Pump: A pump that delivers the same volume of fluid with each revolution regardless of speed.

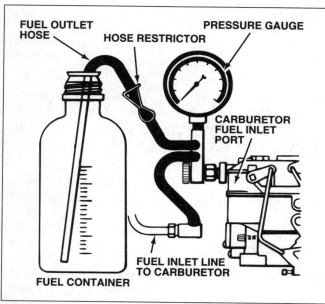

5-2 Setup for performing a fuel pump volume test.

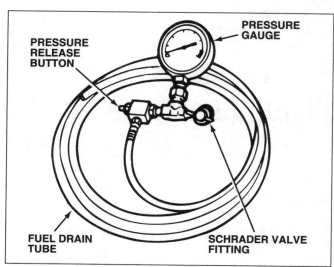

5-3 Fuel injection system pressure testing requires a high-pressure gauge.

Most mechanical fuel pumps can deliver 1 quart or liter of fuel in one minute or less.

Low volume is often caused by a restriction somewhere in the supply line. Check the fuel filter and replace if needed. Also, look for bent, kinked, dented, or leaking fuel lines and hoses. If you suspect that the fuel pump pickup is clogged by contamination in the fuel tank, clear the pickup by applying a short blast of low-pressure compressed air through the fuel-feed line, then repeat the volume test.

Other causes of low fuel pump volume include a worn or broken pump **eccentric**, pushrod, rocker arm, or linkage. Replace the pushrod or pump, then repeat the test. If volume is still low, the eccentric is worn and the camshaft needs replacement.

Electric Pump Testing

Pressure and volume test procedures are identical for a carburetted engine, whether it has a mechanical or electric fuel pump. If testing the pump with the engine not running, bypass the pump control wiring and supply battery voltage directly to the pump. In addition, check the pump electrical circuits for:

- System voltage
- Good ground connections
- All other electrical connections.

The electric pumps used on fuel-injection systems operate at higher pressures, and the test gauge must be able to register pressures that are higher than the system maximum. Typically, a low-pressure, throttle-body, injection system operates on 10 to 15 psi of pressure, while most electronic fuel-injection systems require about 35 psi of fuel pressure. Some gasoline fuel-injection pumps provide more than 50 psi of pressure, and the fuel pressure on a diesel-injection system is even higher. Always check the specifications and use a test gauge that can register more than the maximum, figure 5-3.

Almost all fuel injection systems retain pressure in the fuel lines when the engine is off. As a safety precaution, relieve fuel system pressure before opening fittings to connect the test gauge or make repairs.

Relieving fuel pressure

There are several methods of relieving fuel pressure; follow the recommendations of the vehicle manufacturer. You may have to:

- Apply vacuum to the fuel pressure regulator using a hand-vacuum pump, figure 5-4.
- Attach a special pressure gauge to a **Schrader valve** on the throttle body or fuel rail.
- Remove the fuel pump fuse and run the engine until it dies.
- Remove the wiring harness connector from a fuel injector, then jump one of the injector terminals to ground and apply battery voltage to the other.

Eccentric: Off center or out of round. A shaft lobe which has a center different from that of the shaft.
Schrader Valve: A service valve that uses a spring-loaded pin and internal pressure to seal a port; depressing the pin will open the port.

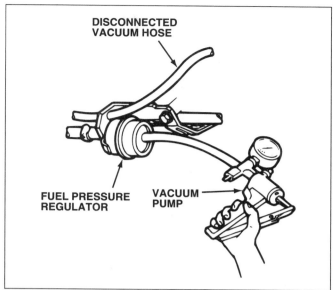

5-4 Relieving residual fuel pressure by applying vacuum to the pressure regulator.

Pressure test

Specifications are usually given for the system-regulated pressure supplied to the injectors. Excess pressure produced by the pump is returned to the fuel tank. A vacuum pump may be required to take readings with and without vacuum applied to the pressure regulator.

Attach the gauge as instructed by the manufacturer to test system pressure. Start the engine and make sure the pressure is within specifications at all speeds and conditions. When specifications for rest, or **residual pressure**, are provided, shut the engine off and leave the gauge attached. Check the gauge reading after waiting the recommended time.

Low system pressure readings may be caused by a faulty fuel pump, accumulator, pressure relief valve, or a clogged fuel filter. Low residual pressure is generally the result of a defective pump check valve.

Volume test

When a pump volume test is specified, it is measured at the fuel return line. Disconnect the return line, attach a piece of fuel hose to the line fitting, and route the open end of the hose into a suitable container. Start the engine and confirm that the specified amount of fuel is delivered within the time allowed.

Fuel System Service

Replacing the fuel filter is generally the only routine service performed on the fuel system. Replace the filter at the recommended interval.

Carbureted engines may have an inline filter installed in the fuel supply line with hose clamps. Some use a filter that installs into the fuel inlet on the carburetor. Remove the supply line fitting with a flare-nut wrench to access the filter.

Fuel injection systems use an inline filter. It may be located in the engine compartment, or underneath the vehicle near the fuel pump. Some injection filters use flare-nut fittings to attach to the lines. Relieve residual pressure before disconnecting the filter.

AIR INDUCTION SYSTEM

The air induction system supplies the engine with a steady stream of fresh air, which is blended with the fuel, to create a combustible mixture. The air induction system consists of:

- Air intake ductwork
- Air cleaner and filter element
- Intake preheat system
- Manifold heat control valves
- Intake manifold.

In addition, the air induction system on some engines includes a turbocharger or supercharger to boost the power output of the engine. The positive crankcase ventilation (PCV) system can also be considered part of the induction system, because the crankcase vapors vent into the intake charge.

Air Intake Ductwork

A typical air intake system is a complex system of ducts, filters, meters, valves, and tubes, figure 5-5. Intake ductwork must be airtight to prevent a loss of air volume to the engine. On a fuel-injected engine, air leakage downstream of the **airflow sensor** can create a lean condition and affect performance. Replace components as needed to maintain system integrity.

Air Filter Service

Carburetted and throttle-body injected engines have the air filter mounted directly to the carburetor or throttle body. To replace the filter element, remove the fasteners holding the housing cover, lift off the cover, and take out the element. Check top and bottom seals for dust leakage and check the filter material for breaks. A

Residual Pressure: A constant pressure held in the fuel system when the pump is not operating.

Airflow Sensor: A sensor used to measure the rate, density, temperature, or volume of air entering the engine.

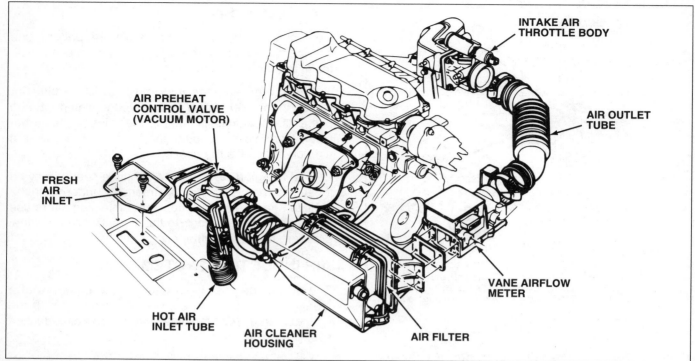

5-5 *Inspect the entire air intake ductwork for signs of leakage and damage.*

paper element cannot be cleaned and must be replaced if it is dirty or torn.

Most fuel-injected engines use a flat filter element installed in a remote air cleaner housing. Some of the ductwork may have to be disconnected and moved aside or removed to access the filter housing. Unfasten the clips holding the housing together, then lift out the element. Replace the element if it is dirty or torn.

Intake Air Preheat Systems

All air cleaners are now designed to control the temperature of the intake air. This speeds warm up and reduces the amount of **hydrocarbon** (HC) and **carbon monoxide** (CO) emissions on a cold running engine. There are three basic tests for these systems:

- Vacuum diaphragm test
- Temperature sensor test
- Thermostatic bulb test.

The vacuum diaphragm and temperature sensor tests are used to check systems with vacuum-actuated hot air control valves, figure 5-6. The third test is for

systems that use the hot air damper controlled by a **thermostatic bulb**, figure 5-7.

Vacuum motor test

Vacuum motor function is tested with a hand-held vacuum pump as follows:

1. Inspect all hoses for correct routing and connections.
2. Note the cold damper position with the engine off. It should cover the hot air passage and allow only cool air into the air cleaner.
3. Disconnect the hose and attach the hand-operated pump to the vacuum motor, figure 5-8.
4. Apply 9 inches of mercury (in-Hg) of vacuum, and the damper should close off the cool air and allow only hot air from the exhaust manifold.
5. With vacuum applied, clamp the line from the vacuum source. The damper should stay closed.

Sensor test

Test the temperature sensor by placing a thermometer inside the air cleaner, near the sensor. Then, start and run the engine while watching the damper door. As soon as the damper begins to open, note the air on the

---

Hydrocarbon: A chemical compound of hydrogen and carbon. A major pollutant from an internal combustion engine. Gasoline itself is a hydrocarbon compound.

Carbon Monoxide: An odorless, colorless, tasteless, poisonous gas. A pollutant produced by an internal combustion engine.

Thermostatic Bulb: A device that automatically responds to changes in temperature to actuate a damper in the air intake passage.

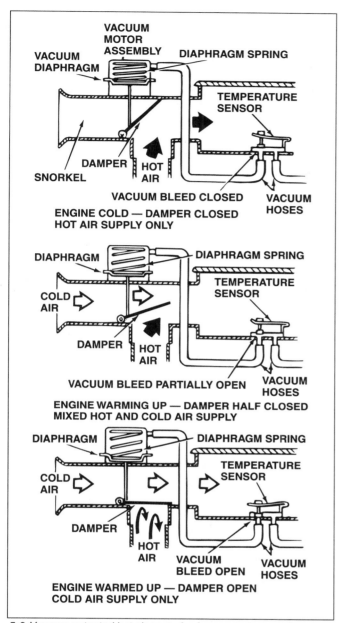

5-6 *Vacuum-actuated hot air control valve uses a temperature-sensitive switch to open and close a blend door.*

thermometer. Continue watching and note the temperature at which the damper door is fully open. If either reading is not within specifications, replace the sensor.

Thermostatic bulb test

A bench test is performed to check the function of a thermostatic bulb, and the entire air cleaner assembly must be removed from the vehicle. To test:

1. Disconnect and remove the intake duct and damper assembly from air cleaner housing.
2. Place the duct assembly in a hot water bath; water temperature should be about 100°F.
3. Allow to soak for 5 minutes, then check the damper position. It should close off the cool air duct.

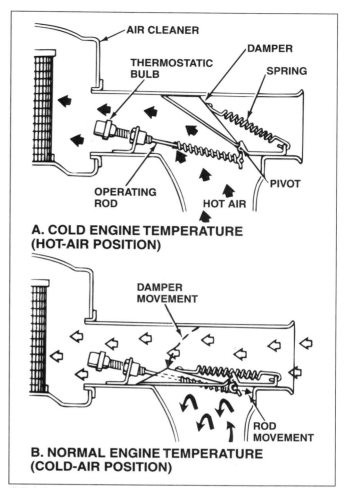

5-7 *A thermostatic bulb responds to temperature changes to open and close a blend door.*

4. Heat the water to maintain a temperature between 130° and 150°F.
5. Soak for 5 minutes and recheck the damper position. It should now close off the hot air inlet. If it does not, the bulb is defective.

Intake Manifold

Intake manifolds seal to the cylinder head with a gasket to provide air-tight passages to the combustion chambers. Engine coolant circulates through most intake manifolds, and the intake manifold on a V-type engine seals to the engine block to seal off the lifter valley. Therefore, manifold to engine sealing must prevent air, coolant, and oil leakage.

Manifold sealing surfaces must be clean, flat, and in good condition. Inspect all of the surfaces before you install the manifolds, and check for warpage using a straightedge and feeler gauge. Be sure the ports and all passageways are free of carbon and any other deposits. Check-fit the gaskets to make sure they will seal properly and all of the openings line up.

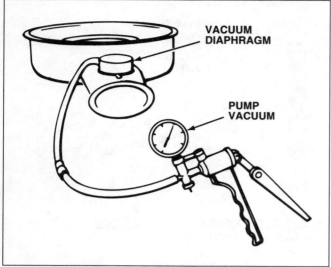

5-8 *Use a handheld vacuum pump to test the diaphragm on a vacuum motor.*

Installation

Intake manifold gaskets normally install dry, without any sealer. Some engines use small, O-ring seals for the coolant passages, in addition to the manifold gasket. If not, a thin bead of RTV silicone around coolant passages prevents coolant leakage. A small dab of silicone sealant is also recommended at the corners of the end seals on a V-type engine to prevent oil leakage. Although the exact procedure will vary by engine, the following installation guidelines apply to any intake manifold:

- Check all of the sealing surfaces on the manifold, cylinder heads, and engine block. Make sure they are clean and free from any old gasket material and sealer.
- Position the manifold on the engine, then install all of the retaining bolts hand-tight. Make sure bolts are in the correct location; several different length bolts may be used.
- Check that all brackets, ground cables, and other attachments that are held in place by the manifold bolts are correctly positioned.
- Tighten bolts to recommended torque in sequence as specified by the manufacturer. Tighten the bolts evenly in stages.

Turbocharger Service

A **turbocharger** is an exhaust-driven compressor used to increase the power output of an engine. The system requires no maintenance other than more frequent

LACK OF POWER OR EMITS BLACK EXHAUST SMOKE	
POSSIBLE CAUSE	**CORRECTION**
Damaged or disconnected air cleaner ducting	Inspect and correct
Restricted air filter element	Replace air filter
Intake or exhaust manifold leak	Check turbocharger installation for air or exhaust leak
Turbocharger damage	Check turbocharger rotating assembly for binding or dragging
Exhaust system restriction	Check exhaust system; check manifold heat control valve
Carburetor or fuel injection problem	Inspect and correct
Incorrect ignition timing, advance–or other ignition problem	Inspect and correct
EGR problem	Inspect and correct
Lack of compression or other engine wear	Inspect and correct
Low boost pressure	Check boost pressure and wastegate operation; adjust or replace as required

5-9

engine oil and filter changes. Turbocharger malfunctions usually fall into one of the following categories:

1. Lack of proper lubrication—caused by the wrong type of oil, a restricted oil supply line, or a worn engine that develops low oil pressure.
2. Dirty or contaminated oil system—caused by infrequent oil changes, or failure of an engine bearing, piston ring, or other internal component, or an oil filter bypass valve stuck open.
3. Contamination in the air intake or exhaust systems—caused by a leaking air duct or missing air cleaner, or a damaged catalytic converter (when installed between the turbocharger and the exhaust manifold).

Troubleshooting

Turbocharger problems are generally revealed by: a lack of power or heavy black smoke from the tailpipe, figure 5-9, high oil consumption or blue smoke from the tailpipe, figure 5-10, or abnormal sounds, figure 5-11. Use the troubleshooting charts to isolate the cause of turbocharger malfunctions.

Inspection and service

Perform a preliminary turbocharger inspection as follows:

1. Check that the air cleaner is in place and that there are no loose connections or leaks, restrictions, or broken ducting.

Turbocharger: A compressor device that uses exhaust gases to turn a turbine that forces the air-fuel mixture into the cylinders.

HIGH OIL CONSUMPTION OR EMITS BLUE EXHAUST SMOKE	
POSSIBLE CAUSE	CORRECTION
Excessive blowby or PCV problem	Check for engine wear; service PCV system
Engine oil leakage	Inspect and correct
Worn engine rings, cylinders, valves, valve guides	Check for engine wear; check for low compression and cylinder leakage
Leaking turbocharger oil seals*	Replace turbocharger
Restricted turbocharger oil return or too much oil in center housing	Check oil return line for restrictions and blow out with compressed air; check center housing for sludge; clean as required
Carburetor or fuel injection problem	Inspect and correct
Incorrect ignition timing, advance, or other ignition problem	Inspect and correct
EGR problem	Inspect and correct

*Smoke and detonation indicates a leak on the compressor side; smoke alone indicates a leak on the turbine side.

5-10

TURBOCHARGER PROBLEMS	
SOUND	CAUSE
Louder than normal noise that includes hissing	Exhaust leak
Uneven sound that changes in pitch	Restricted air intake from a clogged air cleaner filter, bent air ducting, or dirt on the compressor blades
Higher than normal pitch sound	Intake air leak
Sudden noise reduction, with smoke and oil leakage	Turbocharger failure
Uneven noise and vibration	Possible shaft damage, damaged compressor or turbine wheel blades
Grinding or rubbing sounds	Shaft or bearing damage, misaligned compressor or turbine wheel
Rattling sound	Loose exhaust pipe or outlet elbow, damaged wastegate

5-11

under load. Repeat several times and compare results to specifications.

Checking the wastegate actuator

The wastegate is a pressure relief device that protects the system from overcharging. Test the wastegate actuator as follows:

1. Connect a hand-operated pressure pump and gauge to the actuator.
2. Apply about 5 psi of pressure. If pressure drops below 2 psi after one minute, the actuator diaphragm is leaking.
3. Clamp or mount a dial indicator on the turbocharger housing so the plunger contacts the actuator rod.
4. Apply the specified boost pressure to the actuator diaphragm and note the amount of rod movement shown on the indicator. If not within specifications, generally less than 0.015 inch, repair or replace the actuator.
5. Remove the fastener that holds the rod to the wastegate arm or link, then move the arm. It should travel freely through a 45-degree arc. If not, replace the wastegate.

Supercharger Service

A **supercharger** is a crankshaft-driven, positive-displacement pump that supplies an excess volume of

2. Look over the exhaust system for burned areas, leakage, and loose connections.
3. Disconnect the turbocharger air inlet and exhaust outlet and inspect the turbine and compressor with a flashlight and a mirror. Look for bent or broken blades and wear marks on the blades or housing.
4. Turn the wheel by hand. Listen and feel for binding or rubbing. Move both ends of the shaft up and down. There should be little, if any, movement.
5. Check the shaft seal areas, figure 5-12, for signs of oil leakage.

Turbocharger service consists of testing the **boost pressure** and checking the **wastegate** operation. A turbocharger can be disassembled, internal components replaced, and clearances set with a dial indicator. However, common practice is to simply replace a defective unit.

Checking turbocharger boost pressure

Boost pressure can be checked during a road test, or on a chassis dynamometer, using a pressure gauge. Connect the gauge to a pressure port on the compressor side of the turbo, or tee it into the line running to the warning lamp pressure switch. Test with the engine at normal operating temperature. Accelerate from zero to 40 or 50 mph and note the gauge reading

Boost Pressure: The amount of air pressure increase above atmospheric pressure provided by a turbocharger.

Wastegate: A diaphragm-actuated bypass valve used to limit turbocharger boost pressure by limiting the speed of the exhaust turbine.

Supercharger: A crankshaft-driven compressor that delivers an air-fuel mixture to the engine cylinders at a pressure greater than atmospheric pressure.

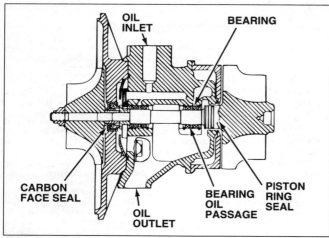

5-12 *Check turbocharger shaft oil seals for leakage.*

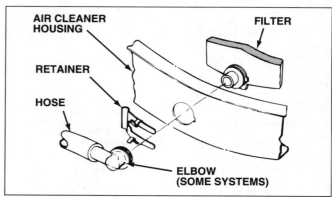

5-13 *Inspect PCV inlet air filters and replace as needed.*

intake air to the engine. The supercharger boosts the pressure and density of the intake air charge to increase engine output. A typical supercharger consists of two lobed rotor shafts supported by bearings in a cast housing. The two shafts are geared together, so they rotate in opposite directions. One of the shafts is driven off the crankshaft by a belt.

The supercharger is supplying boost whenever the engine is running. Most units use a vacuum-operated bypass valve to bleed off excess boost pressure during idle and low-speed operation.

With the exception of belt inspection and adjustment, a supercharger does not require maintenance. Bearings are lubricated by a self-contained oiling system. Some systems allow you to check the oil level; a low oil level indicates an internal problem. Typical problems include:

- Incorrect boost
- Poor response and fuel economy
- Excessive noise
- Oil leakage.

Many problems are a result of a malfunctioning boost control or bypass valve. Valve operation can quickly be checked using a hand-operated vacuum pump. On some systems, you can check boost using a pressure gauge. Boost pressure will vary from about 3.5 to 11 psi, depending on test speed; check specifications for the unit being tested. For the most part, internal components are not serviceable. If a problem is detected, replace the assembly.

Positive Crankcase Ventilation (PCV) System

The PCV system is relatively simple, and most problems are the result of built-up deposits that restrict or prevent air flow. A blocked PCV system may cause:

- Increased oil consumption
- Diluted or dirty oil from sludge, water, or acid in the crankcase
- Escaping blowby vapors from the dipstick tube, oil fill tube, valve-cover gasket areas, and other openings to the crankcase
- Uneven or rich fuel mixture at idle and low speed.

Inspection and testing

Service the PCV system by looking at it, operating it, and replacing defective parts. To inspect:

1. Check all hoses for proper connections, cracks, and clogging.
2. Remove the air cleaner cover and check filter element. Crankcase blowby can clog the filter with oil.
3. Check the crankcase inlet air filter, if equipped. The filter may be in the air cleaner or air hose, figure 5-13, or in the valve cover or oil fill cap.
4. Check for deposits that clog the passages in the manifold.

Vacuum test

Several testers are available to check overall PCV system operation. Some measure pressure in the crankcase. Others test the flow rate at engine idle speed.

When a PCV tester is not available, hold a small piece of stiff paper over the oil fill opening with the engine idling. Crankcase vacuum should pull the paper down against the opening if the system is working properly. If not, or if the paper is blown upward, the system is not working right.

Engine speed drop test

Use the following method to test the PCV system when you do not have a PCV tester, or when the results of the vacuum test are not conclusive:

1. Connect a tachometer and start the engine.
2. Disconnect the PCV valve and the line from the crankcase.

3. Place a finger over the valve and watch the tachometer.
4. Engine speed should drop by 40 rpm or more if the system is working right.

PCV valve test

As a general rule, PCV valves are replaced at recommended service intervals. Replace the valve more often if it fails any test or inspection check. To test a PCV valve:

1. Disconnect the PCV valve from the valve cover or manifold.
2. Run the engine at idle. The valve should make a hissing noise as air passes through it.
3. Put a finger over the end of the PCV valve. You should feel a strong vacuum. If not, check for a clogged valve or restricted hose.
4. Shut off the engine and remove the valve. Shake the valve and listen for the rattle or clicking of the needle in the valve. If not, the valve is bad.

PCV system service

Service of the PCV system usually consists of cleaning or replacing the system air filter, or replacing the connecting hoses or the valve itself.

Filter replacement

On some engines, the PCV inlet air is filtered through the engine air filter. Simply replace the element at the recommended interval. The same is true of separate polyurethane foam filter elements that mount in the air cleaner housing. If a wire screen filter is used, clean it in solvent and allow it to dry.

Filters installed in the oil filler cap are usually made of wire mesh. Remove the filler cap and soak the complete cap and filter in solvent. Allow it to drain and dry in the air. Do not use compressed air to dry these filters because the air pressure will damage the wire mesh.

Hose replacement

Any damaged or deteriorated hoses must be replaced to ensure proper system operation. Use only hoses designed for PCV and fuel system applications. Standard heater hoses will not withstand the blowby vapors.

PCV valve replacement

A PCV valve cannot be distinguished by its appearance; internal valve characteristics are specifically calibrated for each application. Always refer to the part number when you replace a valve. To replace a valve,

simply install the new part in place of the old one. Make sure the valve is installed with the arrow indicating the direction of flow pointing toward the intake manifold. If the valve is mounted in a rubber grommet, make sure it is a snug fit. If the grommet is hardened or cracked, replace it.

ELECTRICAL SYSTEM INSPECTION AND SERVICE

The field of engine repair includes tasks that would normally be considered part of the electrical system. The following discussion provides general guidelines for evaluating, removing, replacing, and servicing the battery and starter motor. For more detailed information, see book six of this series.

Battery Service

When working near or servicing a battery, follow these safety rules:

- Keep area clear of sparks, smoking materials, and open flame.
- Operate charging equipment only in well-ventilated areas.
- Never short across the battery terminals or cable connectors.
- Never connect or disconnect charger leads while the charger is on.
- Remove watches, rings, or other jewelry before working on or near the battery or electrical systems.
- Use a battery carrier to lift and carry batteries.

Inspection

The **electrolyte** level should be above the tops of the plates or at the indicated level within the cells. Add distilled or mineral-free water to refill. Inspect the battery case, terminals, connectors, and holddown tray for rust, corrosion, and damage. Check the battery for cracks, loose terminal posts, or other damage, and replace the battery if defective.

Clean the battery with a solution of baking soda and water. Be careful to keep corrosion off painted surfaces and rubber parts. Rinse the battery and cable connections with fresh water. Disconnect and clean the terminal connections.

For post-type terminals, use connector-spreading pliers to open the connector, then clean the inside of the connectors and the posts with a brush or reamer.

Electrolyte: The chemical solution, generally sulphuric acid and water, in a battery that conducts electricity and reacts with the plate materials.

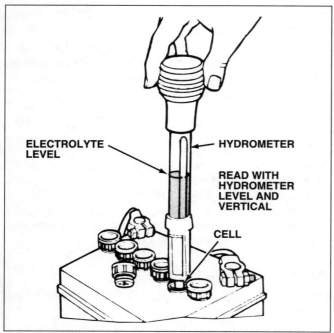

ELECTROLYTE
LEVEL

HYDROMETER

READ WITH
HYDROMETER
LEVEL AND
VERTICAL

CELL

5-14 *Check the specific gravity of a battery with a hydrometer.*

ELECTROLYTE SPECIFIC GRAVITY	BATTERY STATE OF CHARGE
1.265	100%
1.225	75%
1.190	50%
1.155	25%
1.120	Discharged

5-15 *Battery state of charge is determined by the specific gravity of the electrolyte.*

On side-terminal batteries, remove the bolts and clean the threads and contact surfaces with a wire brush. Inspect for damage and corrosion; replace the terminal ends or the cables as required.

State of charge testing

If the battery has removable vent caps, the state of charge can be checked using a hydrometer to test **specific gravity**. Remove the cell caps, insert the hydrometer into a cell, and draw electrolyte into the tester, figure 5-14. Gently shake or tap the hydrometer to keep the float from touching the sides of the tube. Hold the hydrometer at eye level and read specific gravity on the indicator. Return the electrolyte to the cell, then check the remaining cells.

Specific gravity readings for a fully charged battery should be 1.265, figure 5-15. Specific gravity is based on an electrolyte temperature of 80°F. Four points (0.004) should be added or subtracted to the readings for each 10°F difference in temperature. If corrected readings are below 1.225, recharge the battery. The battery should be replaced if readings vary by more than 50 points (0.050) between cells.

Battery charging

Batteries can be charged at rates from 3 amperes (slow charging) to 50 amperes (fast charging). Generally, any battery in good condition can be charged at any current rate if electrolyte gassing and spewing do not occur, and the electrolyte temperature stays below 125°F. Never fast charge a battery that is sulfated or has plate or separator damage. If time allows, a slow rate of 5 to 15 amperes is preferable.

To charge a battery, the electrolyte level should be 0.25 inch above the separators in each cell. Leave the cell caps in place during charging, but be sure the vent holes are open. When charging a sealed, maintenance-free battery, follow the directions of the battery manufacturer for charging rate and time. Charge a battery as follows:

1. Disconnect both battery cables and connect the charger cables to the battery with the correct polarity.
2. Set the charger for the desired rate, then switch it on.
3. Periodically check the specific gravity, electrolyte temperature, and voltage across the battery terminals during charging.
4. If voltage rises above 15.5 volts at any time, lower the charging rate.
5. The battery is fully charged when all cells are gassing freely and specific gravity does not increase for three hours.

After charging, wash and dry the battery top to remove any acid from electrolyte gassing. Check the electrolyte level; replenish if needed. Reconnect the positive cable first, then the ground cable.

Cranking System Service

The cranking system includes the battery, ignition switch, safety switch, starter relay, solenoid, and starter motor, as well as the circuitry that links everything together. A failure at any point of the system can prevent the engine from starting.

Troubleshooting

When an engine fails to crank, perform these preliminary checks:

Specific Gravity: The weight of a volume of liquid divided by the weight of the same volume of water at a given temperature and pressure. Water has a specific gravity of 1.000.

STARTING SYSTEM TROUBLESHOOTING TABLE		
SYMPTOM	**POSSIBLE CAUSE**	**CURE**
• Nothing happens when ignition switch is turned to START.	1. Battery discharged 2. Open in control circuit 3. Defective starter relay or solenoid 4. Open starter motor internal ground	1. Recharge or replace. 2. Check control circuit continuity, repair or replace components as necessary. 3. Replace relay or solenoid. 4. Replace starter.
• Solenoid contacts click or chatter but starter does not operate, OR moveable pole shoe starter chatters or disengages before engine has started.	1. Battery discharged 2. High resistance in system 3. Open in solenoid or moveable pole shoe hold-in winding 4. Defective starter	1. Recharge or replace. 2. Test voltage drop, replace components as necessary. 3. Replace solenoid or replace moveable pole shoe starter. 4. Replace starter.
• Starter motor operates but does not turn engine.	1. Defective starter drive assembly 2. Defective ring gear	1. Replace starter drive. 2. Replace ring gear.
• Starter motor turns the engine over slowly or unevenly.	1. Battery discharged 2. High resistance in system 3. Defective starter 4. Defective ring gear 5. Poor flywheel to starter engagement	1. Recharge or replace. 2. Test voltage drop, replace components as necessary. 3. Replace starter motor. 4. Replace ring gear. 5. Adjust starter position if possible.
• Engine starts but motor drive assembly does not disengage.	1. Defective drive assembly 2. Poor flywheel to starter engagement 3. Shorted solenoid windings 4. Shorted control circuit	1. Replace starter drive. 2. Adjust starter position if possible. 3. Replace solenoid. 4. Test circuit, repair and replace components as needed.

5-16

- Inspect the battery—check for loose or corroded terminals and cable connections, frayed cables, and damaged insulation.
- Inspect the ignition switch—check for loose mounting, damaged wiring, sticking contacts, and loose connections.
- Inspect the safety switch—check for proper adjustment, loose mounting and connections, and damaged wiring.
- Inspect the solenoid—check for loose mounting, loose connections, and damaged wiring.
- Inspect the starter motor—check for loose mounting, proper pinion adjustment, and loose or damaged wiring or connections.

Use the troubleshooting table in figure 5-16 to assist in diagnosing the problem and determine the possible cure.

System testing

To test the starter, the battery must be fully charged and in good condition. Most starting system tests are made while the starter motor is cranking the engine. The engine must not start and run during the tests, or the readings will be inaccurate. To keep the engine from starting, bypass the ignition switch with a remote starter switch, and turn the ignition key off. Alternative methods include disabling the fuel pump or the ignition system, or both. Do not crank the starter motor for more than 15 seconds at a time while testing. Allow two minutes between tests for the motor to cool in order to prevent damage.

Cranking voltage test

Connect the positive lead of a voltmeter to the starter terminal that is energized by the ignition switch, and attach the negative lead to a good ground. Crank the engine and observe the voltmeter. Readings should be over 9.6 volts and the starter should crank freely.

Low readings can be caused by a weak battery, high circuit resistance, or low starter pinion rpm due to high mechanical engine resistance. If readings are good but the starter cranks poorly, there is an internal starter problem.

Current draw test

An ammeter is used to perform this test. The test measures the amperage the circuit requires to crank the engine. Starter current draw specifications are provided by the vehicle manufacturer.

With an inductive ammeter, simply clamp the inductive pickup to the positive battery cable and crank the

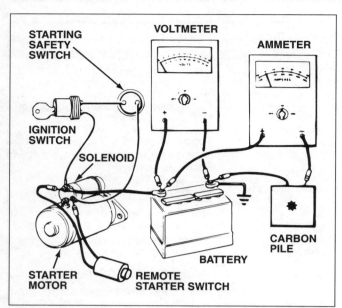

5-17 *Test meter hook up for performing a starter motor current draw test.*

engine for 15 seconds while observing the meter. Compare readings to specifications.

For an ammeter that has mechanical hookups, a voltmeter and carbon pile are required. Test as follows:

1. Refer to figure 5-17 for test connections.
2. Set the carbon pile to its maximum resistance (open).
3. Crank the engine and watch the voltmeter reading.
4. With the starter motor off, adjust the carbon pile until the voltmeter reading matches the reading taken in step 3.
5. Note the ammeter reading and set the carbon pile back to open. Then, compare the ammeter reading to specifications.

Interpret test results as described for the cranking voltage test.

Circuit resistance tests

The starting system is composed of three circuits:

- Insulated circuit—all high current cables and connections between the battery and starter
- Ground circuit—the return path from the starter motor to the battery
- Control circuit—the low current wiring, switches, and relays used to energize the starter motor.

Resistance tests on all three circuits can be performed with a voltmeter. One voltmeter lead is connected to the

battery terminal and the other to various connections in the circuit. Manufacturers recommend different specific test points for their vehicles. The engine is cranked and the voltmeter observed. As a general rule, the maximum drop should not exceed 0.1 volt per connection.

Starter motor removal and replacement

You may have to remove, loosen, or relocate heat shields, support brackets, or exhaust pipes to access the starter motor. Always begin by disconnecting the battery ground cable at the battery.

After gaining access to the starter motor, disconnect all of the wires from the starter motor or solenoid. Remove the mounting bolts securing the starter motor to the engine, then remove the starter motor.

Once the starter is removed, inspect the ring gear. Look for any chipped, broken, missing, or excessively worn teeth. Turn the engine over by hand to examine the entire ring gear. If damage is found, replace the ring gear.

Bolt the new starter motor to the engine. Check the flywheel/starter engagement. Some designs require shims to provide correct engagement of the starter pinion with the flywheel. Add or remove shims until the engagement is correct. Connect all wires to the solenoid or motor terminals and replace all items removed on disassembly.

IGNITION SYSTEM INSPECTION AND SERVICE

Ignition system service is limited to component inspection, replacement, and adjustment in this book. For more specific information on ignition system diagnosis, testing, and repair, refer to book eight of this series.

Primary Wiring and Connectors

Primary wiring and connectors are a potential source of high **resistance**, as well as open or grounded circuits that can prevent an engine from starting. This type of problem can be found with simple voltmeter and ohmmeter tests. Check for correct supply voltage first. If voltage is present, check for a **voltage drop** using a voltmeter. When voltage drop indicates high or low resistance, verify with an ohmmeter. Many problems can easily be solved by cleaning connectors or repairing wiring.

Resistance: Opposition to electrical current.

Voltage Drop: The measurement of the loss of voltage caused by the resistance of a conductor or a circuit device.

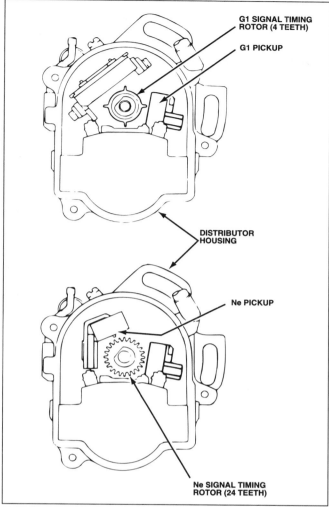

G1 SIGNAL TIMING ROTOR (4 TEETH)

G1 PICKUP

DISTRIBUTOR HOUSING

Ne PICKUP

Ne SIGNAL TIMING ROTOR (24 TEETH)

5-18 *Some distributors use two pickup coils. This Toyota distributor provides two position signals to the PCM.*

Ignition Signaling Devices

Electronic ignition systems use one of several types of **signal generators** to trigger spark plug firing. These sensors tell the ignition module or PCM the speed and position of the crankshaft.

Magnetic pulse generator

These devices produce an alternating current (AC) signal from the movement of a metal tooth past a pickup coil. On many, the air gap between the pickup coil and trigger wheel is adjustable. Some systems couple two generators together for information about crankshaft position and speed, figures 5-18 and 5-19. Gap has no

affect on the dwell period; dwell is determined by the control module. The air gap must be set to a specific clearance when a new pickup unit is installed. During use, the air gap should not change. However, it should be checked before performing troubleshooting tests.

The air gap between the trigger wheel and the pickup coil can change because of worn distributor bushings. With a conventional distributor, the gap can also change because of breaker plate and vacuum advance mechanism wear.

Check and adjust the air gap using a nonmagnetic (brass) feeler gauge, figure 5-20. The pickup coil magnet will attract a steel gauge, causing an inaccurate adjustment. Air gaps differ between manufacturers, as well as between models. You must have the correct specifications. To check and adjust air gap:

1. Rotate the engine to align one trigger wheel tooth with the pickup coil.
2. Place a nonmagnetic feeler gauge of the specified thickness between the wheel tooth and the pickup coil. The feeler gauge should just barely make contact on both sides.
3. If adjustment is necessary, loosen the lockscrew and use a screwdriver blade to shift the pickup plate. Tighten the lockscrew. If force is required to remove the feeler gauge, the adjustment is too tight.
4. Apply vacuum to the advance diaphragm as you watch the breaker plate move through its full travel. Be sure the pickup coil does not hit the trigger wheel tooth. Release vacuum and recheck the air gap. If the pickup plate is loose, the distributor should be overhauled.

Hall-effect switch

Many ignition systems use a **Hall-effect switch**, which produces a digital signal by blocking and exposing a magnetic field to the base of the Hall-effect element, figure 5-21. These devices require a reference voltage to operate and are more accurate than magnetic generators.

A Hall-effect switch requires no maintenance or adjustment, and its condition cannot be determined by a visual inspection. Verify the signal using a digital multimeter, oscilloscope, or logic probe, and replace as necessary.

Signal Generator: An electrical device that creates a voltage pulse that changes frequency or amplitude, or both, in relation to rotational speed.

Hall-effect Switch: A semiconductor that produces a voltage in the presence of a magnetic field. This voltage can be used to cycle a transistor on and off to produce a variable frequency digital signal.

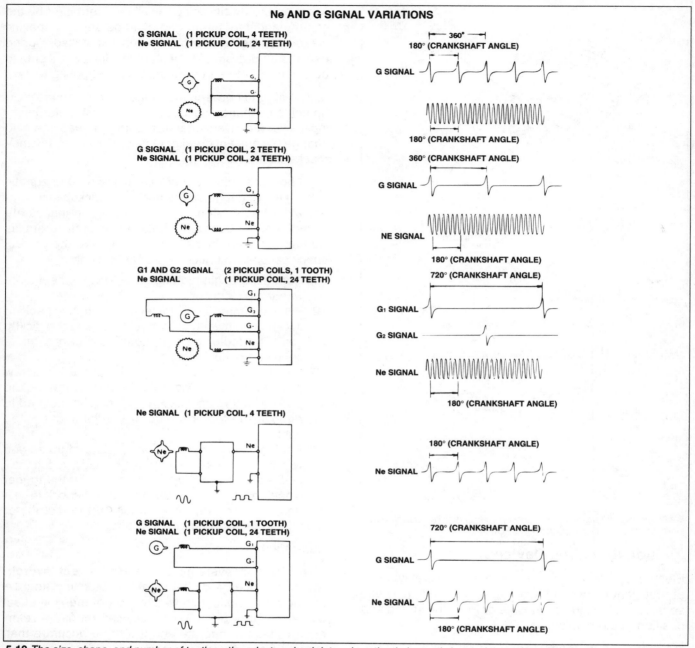

Ne AND G SIGNAL VARIATIONS

5-19 *The size, shape, and number of teeth on the reluctor wheel determines the timing and shape of the signal waveform.*

Optical sensor

An optical sensor uses a light-emitting diode (LED), a shutter wheel, and a phototransistor to produce a digital signal that changes **frequency** in proportion to rotational speed. Like a Hall-effect switch, an optical sensor is maintenance and adjustment free. Check sensor operation with the appropriate test equipment and replace the unit as needed.

Secondary Voltage Check

Often, it is useful, when diagnosing a no-start condition, to verify that secondary voltage is available from the coil by performing a spark test. Begin by disconnecting a spark plug cable from the plug or removing the coil cable at the distributor cap. Hold the end of the disconnected cable about ¼ inch (6 mm) from a good ground, and watch for a spark while cranking the

Frequency: The number of periodic voltage oscillations, or waves, occurring in a given unit of time, usually expressed as cycles per second, or Hertz.

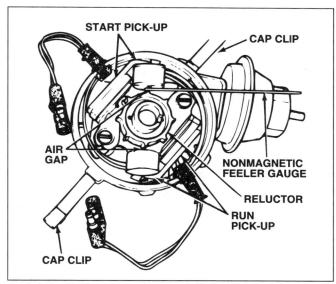

5-20 *Adjusting the reluctor-to-pickup coil air gap on a Chrysler dual-pickup distributor.*

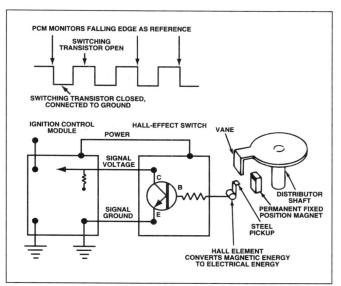

5-21 *A Hall-effect switch requires an external power source to generate a digital signal voltage that changes frequency in proportion to speed.*

engine, figure 5-22. A bright, blue spark should be clearly visible. Be aware: This creates high open-circuit voltage that causes damage on some electronic systems. The preferred method is to use a spark plug simulator, figure 5-23.

Distributor Cap and Rotor Replacement

Never clean cap electrodes or the rotor terminal by filing. If electrodes are burned or damaged, replace the part. Filing changes the rotor air gap and increases resistance. Most caps are replaced by simply undoing spring-type clips, L-shaped lug hooks, or holddown screws. Lift the old cap off and install a new one. When replacing the cap, be sure to install the spark plug cables in correct firing sequence.

Rotors may be retained by holddown screws or may simply slip-fit onto the shaft. Rotors also use one or more locating lugs to correctly position them on the distributor shaft, figure 5-24. Make sure the rotor is fully seated. If not, it can strike the cap when the engine is started.

Use caution when installing caps on distributors with the coil inside the distributor. The carbon brush in the cap may break if the cap is pushed into place after it is seated. Align the components before installing the distributor cap.

Ignition Timing

Ignition timing on older engines with a conventional distributor may be mechanically adjusted. However, the timing on most late-model engines is electronically regulated and cannot be adjusted. Timing advance is also electronically controlled on newer systems,

while a conventional distributor advances timing in relation to engine speed using mechanical weights, a vacuum diaphragm, or a combination of both.

Mechanical ignition timing

Ignition timing can be advanced within the distributor centrifugally and with vacuum. Engine speed governs centrifugal advance; its load, vacuum advance. As speed increases, so does the timing advance. Also, light-load operating conditions will advance timing to decrease fuel consumption.

Dual diaphragm distributor units can either use one diaphragm for advance and the other for retard, or use one for low advance and the other for high advance.

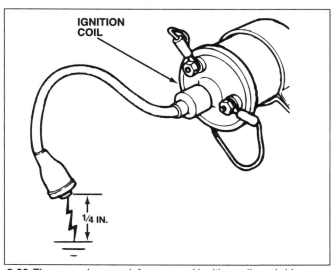

5-22 *The secondary spark from a good ignition coil can bridge a considerable gap.*

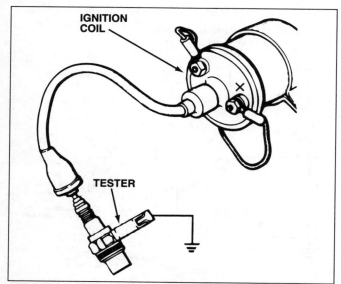

5-23 *Using a spark plug simulator to check for secondary voltage.*

The advance/retard unit uses ported vacuum to advance the timing, and manifold vacuum to allow the timing to retard further in the absence of ported vacuum.

The advance/advance unit uses manifold vacuum through a thermal vacuum switch to advance the timing on cold engines for better driveability. **Ported vacuum** is typically used to advance the timing during driving under load. When testing an ignition system with a dual diaphragm, it is important to know which system you have.

Static timing procedure

The following procedure can be used to statically time an electronic ignition system:

1. Rotate the engine until the crankshaft timing mark aligns with the proper timing specification and number one cylinder is on the compression stroke.
2. Loosen the distributor holddown clamp and bolt.
3. Rotate the distributor housing in the direction of rotor rotation until the trigger wheel tooth is aligned with the pole piece, and the rotor is pointed toward number one plug wire in the distributor cap, figure 5-25.
4. Tighten the holddown bolt.
5. Check and adjust the dynamic timing.

Dynamic timing

Dynamic timing is adjusted by rotating the distributor housing while the engine is running at idle and is at normal operating temperature. Most manufacturers specify that the distributor vacuum lines be disconnected and plugged. Precise spark timing is important

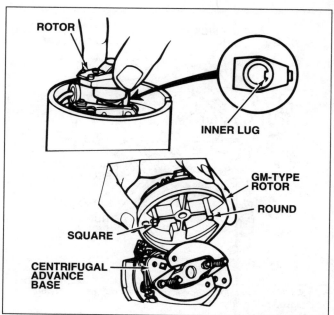

5-24 *Alignment lugs hold the rotor in position on the distributor shaft.*

to ensure correct engine performance and to reduce vehicle emissions.

1. Connect a timing light and tachometer to the engine.
2. Disconnect and plug distributor vacuum lines to prevent unwanted vacuum advance.
3. Slightly loosen the distributor holddown bolt.
4. Start the engine. Adjust idle speed to specifications to prevent unwanted centrifugal advance.
5. Point the timing light toward the timing marks. Carefully sight the marks at a right angle, figure 5-26. Looking at the marks from an incorrect position can result in an initial timing error of as much as 2 degrees.

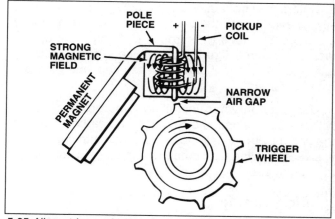

5-25 *Align a trigger wheel tooth with the pole piece when static timing an engine that uses a pickup coil.*

Ported Vacuum: The low-pressure area, or vacuum, located just above the intake air throttle plates of an engine.

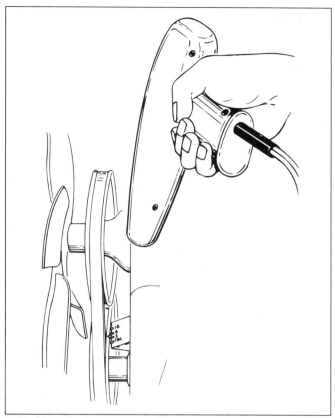

5-26 *Observing timing marks with a stroboscopic timing light.*

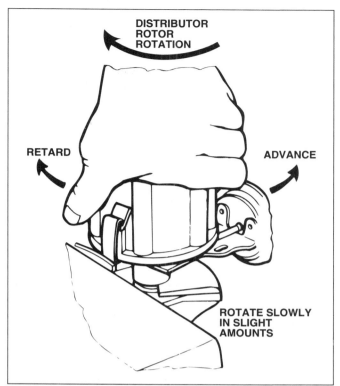

5-27 *Slowly rotate the body of a conventional distributor to adjust ignition timing.*

6. Rotate the distributor housing until the proper marks are aligned, figure 5-27.
7. When the timing is set to specifications, carefully tighten the distributor holddown bolt. Recheck the timing.

Electronic timing control

The ignition module or power control module (PCM) calculates ignition timing on most late-model vehicles. The PCM receives inputs from sensors giving engine temperature, rpm, load, vehicle speed, throttle position, amount of EGR, barometric pressure, and possible detonation.

Typically, setting base timing involves disabling the computer from controlling timing. Follow the procedure on the engine decal, or check other sources if the decal does not specify. This operation can be done by disconnecting a wire or coolant sensor, jumping between two wires, or jumping the connectors in the **data link connector** (DLC). It is important to follow the correct procedure.

Spark Plug Replacement

Spark plug access is limited on many late-model engines, due to a maze of air conditioning and emission control plumbing and engine-driven accessory mounting. Engine accessories may have to be loosened from their mountings and moved to get to the plugs. Air conditioning compressors, air pumps, and power steering pumps are frequent candidates for relocation during plug service. Whenever you must move one of these accessories, be careful of its plumbing and wiring. The spark plugs on some engines are most easily reached from underneath the engine.

Spark plug removal

To remove the spark plugs:

1. Disconnect cables at the plug by grasping the boot and twisting gently while pulling. Do not pull on the cable. Insulated spark plug pliers, figure 5-28, will provide a better grip and are recommended when working near hot manifolds.
2. Loosen each plug one or two turns with a spark plug socket, then blow dirt away from around the plugs with compressed air.

Data Link Connector: An electrical connector that allows a scan tool to communicate with the powertrain control module (PCM).

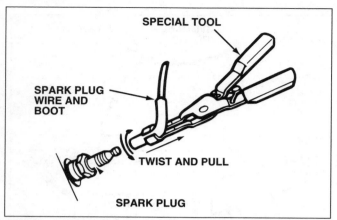

5-28 *Insulated spark plug cable pliers prevent shock and protect cables.*

3. Remove the plugs, keeping them in cylinder number order for inspection.
4. When removing gasketed plugs, be sure the old gasket comes out with the plug.

Spark plug installation

Spark plugs, both new and used, must be correctly gapped before they are installed. Although a wide variety of gapping tools are available, a round wire feeler gauge is the most efficient for used plugs, figure 5-29. A flat feeler gauge should not be used, because the measurement will be inaccurate. Adjust the gap by carefully bending the ground electrode, figure 5-30. Set the gap to specifications provided by the engine manufacturer:

- Do not assume that new plugs are correctly gapped.
- Do not try to set a wide-gap plug (electronic ignition) to a small-gap specification. This will damage the electrode.
- Do not try to set a narrow-gap plug (breaker-point ignition) to a wide-gap specification. This will result in misaligned electrodes.
- Do not make gap adjustments by tapping the electrode on a workbench or other solid object. This can cause internal plug damage.

Cleaning the threaded plug holes in the cylinder head with a thread chaser will ensure easy spark plug installation. With aluminum heads, use the tool carefully to avoid damaging the threads.

Some manufacturers recommend using an antiseize compound or thread lubricant on the plug threads. Thread lubricant recommendations are rare. Use only when specified by the manufacturer. Antiseize compound is commonly used when installing spark plugs in aluminum cylinder heads. Use the specific compound recommended by the manufacturer. Not all are

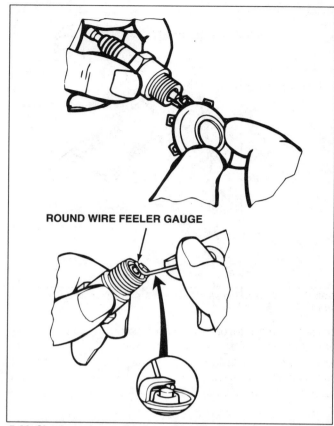

5-29 *Check spark gap with a round wire feeler gauge.*

compatible with aluminum. Whenever thread lubricant or antiseize is used, reduce the tightening torque slightly.

Once the plug gap has been properly set, install as follows:

1. Wipe any dirt and grease from the cylinder head plug seats with a clean cloth.
2. Check that gaskets used on the plugs are in good condition and properly installed.
3. Install the plugs in the engine and run them in by hand.
4. Tighten the plugs to specification with a torque wrench. General torque values are listed in figure 5-31.

EXHAUST MANIFOLD INSPECTION AND SERVICE

Before installing an exhaust manifold, check the sealing surfaces, gasket fit, and hardware condition. Look over the mating surface on the cylinder head for any problems such as pulled or damaged threads.

Carefully inspect the manifold for signs of cracking or other damage. Magnetic particle testing, as explained in chapter two, works well for locating hairline

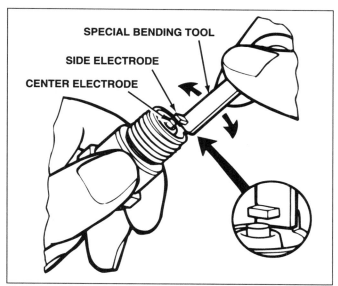

5-30 *Carefully bend the electrode with a special tool to adjust spark plug gap.*

cracks on a cast-iron manifold. Check the straightness of the manifold flanges. A quick, easy way to do this is to run a file straight across each flange. If flat, the file will scuff the entire surface area. You can take the surface down with the file if the flange is slightly warped.

Exhaust manifolds seal to the head with either a single gasket, an individual gasket for each port, or a combination intake/exhaust gasket. Some gaskets are stainless steel on one side and a composition material on the other. These are installed with the steel side toward the manifold.

PLUG TYPE	CAST-IRON HEAD		ALUMINUM HEAD	
	Foot-Pounds	Newton-Meters	Foot-Pounds	Newton-Meters
14-MM Gasketed	25-30	34-40	13-22	18-30
14-MM Tapered Seat	7-15	9-20	7-15	9-20
18-MM Tapered Seat	15-20	20-27	15-20	20-27

5-31 *Always tighten spark plugs to specified torque.*

Replace all exhaust manifold mounting hardware, studs, nuts, and bolts with new pieces. The extreme temperatures these fasteners are exposed to result in corrosion and a loss of tensile strength. Over time, the metal becomes brittle and will break easily. Always use brass or copper-plated nuts. These will not bind on the threads as the metals expand and contract.

Once everything passes inspection, install the exhaust manifold and gasket. Start all of the nuts, or bolts, and run them up hand-tight. Be sure any brackets or shrouds that attach with the manifold are in place. Then bring the fasteners up to specified torque in stages.

1. When tested for one minute at 500 rpm, most mechanical fuel pumps deliver:
   a. 1 quart of fuel
   b. 1 pint of fuel
   c. 1 gallon of fuel
   d. 1 cup of fuel

2. A typical electronic multi-port injection system operates at a system pressure of about:
   a. 25 psi
   b. 35 psi
   c. 45 psi
   d. Over 50 psi

3. Low system pressure on an engine with electronic fuel injection can be caused by all of the following *EXCEPT*:
   a. Defective fuel pump
   b. Faulty fuel pressure regulator
   c. Clogged fuel filter
   d. Defective fuel pump check valve

4. The thermostatically controlled air cleaner should:
   a. Close the hot air passage during warmup
   b. Open the hot air passage during warmup
   c. Provide hot air at all times
   d. Provide cool air at all times

5. Which of the following tests is *NOT* used to check the operation of an intake air preheat system?
   a. Vacuum diaphragm test
   b. Temperature sensor test
   c. A manifold vacuum test
   d. Thermostatic bulb test

6. Most turbocharger failures are caused by:
   a. Excessive high-speed operation
   b. Exhaust temperatures too high
   c. Dirt and contamination
   d. Operation with leaded gasoline

7. To check the boost pressure of a turbocharger, a pressure gauge is installed:
   a. On the compressor side of the turbocharger
   b. On the turbine side of the turbocharger
   c. At the wastegate port of the turbocharger
   d. In the intake air plenum

8. To bleed off excess boost pressure during idle and low speed operation, most supercharger engines use:
   a. A vacuum-operated bypass valve
   b. A pressure-controlled wastegate
   c. An electric cut-out switch
   d. A pop-off valve

9. A clogged PCV system can cause all of the following *EXCEPT*:
   a. Increased oil consumption
   b. Lean fuel mixture
   c. Oil dilution
   d. Oil leaks

10. When doing an engine speed drop test of the PCV system, the engine idle speed should drop by at least:
    a. 20 rpm
    b. 40 rpm
    c. 60 rpm
    d. 80 rpm

11. The specific gravity of a fully charged battery with an electrolyte temperature of 80°F will be about:
    a. 1.225
    b. 1.245
    c. 1.265
    d. 1.285

12. The preferred rate for charging an automotive battery is:
    a. Below 5 amps
    b. 5 to 15 amps
    c. 15 to 25 amps
    d. Over 25 amps

13. Low readings on a starter motor cranking voltage test can be caused by all of the following *EXCEPT*:
    a. Weak battery
    b. High circuit resistance
    c. Low starter pinion speed
    d. Internal starter problem

14. Which of the following is *NOT* one of the starting system circuits?
    a. Control circuit
    b. Isolated circuit
    c. Insulated circuit
    d. Ground circuit

15. Starter circuit resistance tests are performed with a(n):
    a. Ammeter
    b. Oscilloscope
    c. Ohmmeter
    d. Voltmeter

16. Which of the following signal generating devices produces an alternating current (AC) signal to let the PCM know how fast the crankshaft is turning?
    a. Hall-effect switch
    b. Magnetic pulse generator
    c. Optical sensor
    d. All of the above

17. To check base timing on a late-model computer-controlled engine, it may be necessary to:
    a. Disconnect the oxygen sensor
    b. Bypass the idle air control valve
    c. Disable computer controlled timing
    d. Disconnect the throttle position sensor

18. On a conventional distributor, the air gap between the trigger wheel and the pickup coil can change because of:
    a. Incorrect base timing setting
    b. Incorrect timing advance
    c. Worn distributor bushings
    d. Worn distributor cap and rotor

19. Spark plug gap is checked using a:
    a. Round wire feeler gauge
    b. Flat blade feeler gauge
    c. Non-magnetic feeler gauge
    d. All of the above

20. How do exhaust manifold gaskets that are stainless steel on one side and a composition material on the other install?
    a. With the steel side toward the cylinder head
    b. With the steel side toward the manifold
    c. Either way
    d. Depends on the engine

This sample test can help you review your knowledge of this entire book. The format of the questions is similar to the certification tests given by the National Institute for Automotive Service Excellence. Generally, the questions here are more difficult than the programmed study questions you answered as you read the technical material in this book.

Read these review questions carefully, then read all the possible answers before making your decision. Always select the **best possible answer**. In some cases, you may think all the answers are partially correct, or you may feel that none is exactly right. But in every case, there is a **best** answer; that is the one you should select.

Answers to the questions in this sample test will be found near the end of this book, following the glossary. If you answer at least 30 of these questions correctly, then you can be confident of your knowledge of the subjects covered in this book and in the ASE Certification Test A1 Engine Repair. If you answer fewer than 30 correctly, you should reread the text and take another look at the illustrations. Also, check the glossary as you review the material.

1. A compression test was performed on a four-cylinder engine with the following results: cylinder #1 = 138 psi, cylinder #2 = 127 psi, cylinder #3 = 115 psi, cylinder # 4 = 135 psi.
Technician A says low compression in cylinders 2 and 3 indicates that the cylinder head gasket is leaking between these two cylinders.
Technician B says the compression test is not conclusive enough to determine the problem. Uneven readings may be caused by burnt, sticking, or improperly adjusted valves or carbon deposits on the valve faces.
Who is right?
a. A only
b. B only
c. Both A and B
d. Neither A nor B

2. A vacuum gauge is connected to an engine. The gauge reads 18 in-Hg at idle. When engine speed is increased, the gauge momentarily drops to near zero, then slowly rises and stabilizes at 9 in-Hg. The likely cause of these readings would be:
a. Weak piston rings
b. Incorrect fuel mixture
c. Restricted exhaust
d. Retarded ignition timing

3. Manifold vacuum, compression, cylinder leakage, and cylinder power balance test results are all within specification, but the engine suffers from high oil consumption, even though the valve guide seals were recently replaced.
Technician A says that the problem may be caused by a clogged PCV valve or worn valve guides.
Technician B says clogged oil drainback passages or worn pushrod tips can increase oil consumption.
Who is right?
a. A only
b. B only
c. Both A and B
d. Neither A nor B

4. A cylinder leakage test is being performed on a 4-cylinder engine. When testing the number 2 cylinder, air escapes from the number 3 spark plug hole.
Technician A says that a leaking head gasket could be the problem.
Technician B says that the number 2 intake valve not seating could be the problem.
Who is right?
a. A only
b. B only
c. Both A and B
d. Neither A nor B

5. A cylinder compression test reveals fairly even compression, but lower than specification. A second wet test results in higher readings on all cylinders. These results indicate:
a. Poor valve seating
b. Worn valve guides
c. Worn piston rings
d. Leaking head gasket

6. The needle of a vacuum gauge rapidly vibrates between 14 and 18 in-Hg with the engine running at idle. As engine speed increases, the gauge stops vibrating and stabilizes. These readings are most likely caused by:
a. Weak piston rings
b. Incorrect fuel mixture
c. Worn valve guides
d. Weak valve springs

7. Which of the following is the proper procedure for performing a cylinder balance test?
a. Run the engine at idle, short the spark plugs one at a time, and note any rpm drop
b. Run the engine at 1500 rpm, short the spark plugs two at a time, and note any rpm drop
c. Run the engine at idle speed, shorting all but two spark plugs, and note any rpm drop
d. Run the engine at 1500 rpm, short the spark plugs one at a time, and note any rpm drop

8. An engine has a hollow, metallic clatter when started cold. The noise diminishes and eventually goes away as the engine warms up to operating temperature.
Technician A says the noise could result from low oil pressure caused by worn main bearings.
Technician B says the noise could be piston slap caused by too much piston-to-cylinder wall clearance.
Who is right?
a. A only
b. B only
c. Both A and B
d. Neither A nor B

9. The needle on a vacuum gauge oscillates back and forth when the engine is run at idle. This generally indicates which of the following?
   a. Late valve timing
   b. Poor air-fuel mixture
   c. A blown cylinder head gasket
   d. An air leak in the intake manifold

10. Technician A says that grinding the valve face will increase valve spring installed height. Grinding the valve tip will correct installed height.
    Technician B says that grinding the valve face will decrease valve spring installed height. Installing shims under the spring will correct installed height.
    Who is right?
    a. A only
    b. B only
    c. Both A and B
    d. Neither A nor B

11. After facing, valves must have a margin of at least?
    a. ⅟₆₄ inch
    b. ⅟₃₂ inch
    c. ⅟₁₆ inch
    d. ⅛ inch

12. Resurfacing an OHC cylinder head will?
    a. Increase valve lift
    b. Alter cam timing
    c. Change valve lash
    d. Lower compression

13. Valve spring installed height is measured from?
    a. Spring seat to valve tip
    b. Valve guide top to valve keeper groove
    c. Valve retainer top to valve stem end
    d. Spring seat to valve retainer

14. Which of the following tools will give the most accurate measurements when calculating valve guide wear?
    a. Split-ball gauge
    b. Inside micrometer
    c. Dial indicator
    d. Dial bore gauge

15. Which of the following methods is **NOT** used for detecting cracks in a cast-iron cylinder head?
    a. Dye penetrant testing
    b. Pressure testing
    c. Magnetic particle testing
    d. Leak-down testing

16. Technician A says that it is a good practice to replace torque-to-yield cylinder head bolts with new ones whenever they are removed. Technician B says that torque-to-yield cylinder head bolts can be reinstalled if you torque them to specification, then tighten them an additional 90 degrees.
    Who is right?
    a. A only
    b. B only
    c. Both A and B
    d. Neither A nor B

17. Technician A says the close wound end of a variable-pitch valve spring should be installed up, with the tight coils at the tip end of the valve.
    Technician B says the close wound end of a variable-pitch valve spring should be installed down, with the tight coils at the spring seat of the cylinder head.
    Who is right?
    a. A only
    b. B only
    c. Both A and B
    d. Neither A nor B

18. A cylinder bore measures 3.067 inch below ring travel at the bottom and 3.077 below the ridge at the top. The standard factory bore is 3.065. What would be the proper repair?
    a. Deglaze the cylinder walls and install the old pistons with new cast-iron rings
    b. Bore and hone the cylinder to 3.097 inch and install 0.030 inch oversize pistons with new rings
    c. Bore and hone the cylinder to 3.075 inch and install knurled pistons with new rings
    d. Bore and hone the cylinder to 3.095 inch and install 0.030-inch oversized pistons with new rings

19. Technician A says a full-floating wrist pin should be checked for clearance fit.
    Technician B says a press-fit wrist pin should be checked for interference fit.
    Who is right?
    a. A only
    b. B only
    c. Both A and B
    d. Neither A nor B

20. Two cylinder bore measurements are taken: one perpendicular (at a right angle) to the crankshaft and the other in line with the crankshaft. These two measurements are used to determine:
    a. Cylinder out-of-round
    b. Cylinder taper
    c. Cylinder warpage
    d. Cylinder ridge

21. When installing a piston assembly, the notch on the head of the piston will generally face toward:
    a. The rear of the engine
    b. The major thrust side
    c. The front of the engine
    d. The minor thrust side

22. Before installing pistons into freshly honed cylinder bores, the cylinders should be cleaned using:
    a. Clean solvent
    b. Hot soapy water
    c. Water soluble oil
    d. Carbon tetrachloride

23. Failure to remove the cylinder ridge before removing the piston and connecting rod assemblies from the engine block can result in:
    a. Scored cylinder walls
    b. Bent connecting rods
    c. Broken piston skirts
    d. Damaged piston ring lands

24. In a high mileage engine, cylinder wear is normally the greatest at the:
    a. Top of the cylinder measured parallel to the crankshaft
    b. Top of the cylinder measured at right angles to the crankshaft
    c. Bottom of the cylinder measured parallel to the crankshaft
    d. Bottom of the cylinder measured at right angles to the crankshaft

25. A diagonal, or angled, wear pattern on the thrust surface of a piston is an indication of:
    a. A bent connecting rod
    b. Too much piston pin clearance
    c. An out-of-round cylinder bore
    d. A collapsed piston skirt

26. During an oil pressure test, low oil pressure is noted only at low engine speeds, and there are no abnormal engine noises. These findings may indicate:
    a. Faulty hydraulic valve lifters
    b. Pressure-oiled rocker arm shaft assembly wear
    c. Badly worn engine bearings
    d. An open oil pump pressure relief valve

27. A cooling system pressure tester is connected to the radiator and pumped up to the proper pressure. The engine is started, and after a while, the pressure begins to rise. What could be the cause of the problem?
    a. The thermostat is bad
    b. The bypass hose is blocked
    c. The tester is being used incorrectly
    d. The head gasket is leaking

28. Which of the following may be a cause of low oil pressure?
    a. Worn piston rings
    b. Excessive valve guide clearance
    c. Excessive main bearing clearance
    d. Stretched timing chain

29. Technician A says that two measurements, end clearance and rotor-to-housing clearance, are taken to evaluate a rotor-type oil pump.
    Technician B says that three measurements: end clearance, gear-to-gear clearance, and gear-to-housing clearance, are taken to evaluate a gear-type oil pump. Who is right?
    a. A only
    b. B only
    c. Both A and B
    d. Neither A nor B

30. High or low resistance in the primary ignition circuit can be located by:
    a. Current draw testing
    b. Checking available voltage
    c. Voltage drop testing
    d. State of charge testing

31. A Hall-Effect switch on a computer-controlled engine triggers spark plug firing through the use of:
    a. A light-emitting diode and a reluctor plate
    b. A semiconductor and a permanent magnet
    c. A pick-up coil and a trigger wheel
    d. A magnetic pulse generator and an armature

32. A voltage loss in the primary circuit may be caused by all of the following **EXCEPT**:
    a. High circuit resistance
    b. Insufficient battery voltage
    c. Low charging system output
    d. Excessive starter motor current draw

33. While the engine is running, a technician pulls the PCV valve out of the valve cover and puts his thumb over the valve opening. There are no changes in engine operation.
    Technician A says the PCV valve could be stuck in the open position.
    Technician B says the hose between the intake manifold and the PCV valve could be plugged. Who is right?
    a. A only
    b. B only
    c. Both A and B
    d. Neither A nor B

34. Residual fuel pressure reads low on an engine with electronic fuel injection. The most likely cause is:
    a. A defective fuel pump
    b. A faulty fuel-pressure regulator
    c. A clogged fuel filter
    d. A defective fuel pump check valve

35. Black exhaust smoke from a turbocharged engine can indicate:
    a. Coolant leakage into the cylinders
    b. Leaking oil seals in the turbo
    c. Clogged intake air filter
    d. High boost pressure

36. Insufficient system pressure in a port fuel-injection system can be caused by all of the following **EXCEPT**:
    a. A defective fuel-pressure regulator
    b. A sticking fuel pump check valve
    c. A clogged fuel filter
    d. A faulty fuel accumulator

37. The thermostatically controlled air cleaner should:
    a. Close hot air passage during warm-up
    b. Open hot air passage during warm-up
    c. Provide hot air at all times
    d. Provide cool air at all times

38. When slow-charging a battery, the electrolyte temperature should not rise above:
    a. 112°F
    b. 115°F
    c. 125°F
    d. 135°F

39. Which of the following circuits is **NOT** part of the cranking system?
    a. Ignition switch
    b. Engine speed sensor
    c. Starter relay
    d. Battery

40. To correct specific gravity readings taken at a temperature of 90°F:
    a. Add 0.004 to the hydrometer reading
    b. Add 0.006 to the hydrometer reading
    c. Add 0.008 to the hydrometer reading
    d. Add 0.010 to the hydrometer reading

Aerobic: Curing in the presence of oxygen.

Airflow Sensor: A sensor used to measure the rate, density, temperature, or volume of air entering the engine.

Anaerobic: Curing in the absence of oxygen.

Atmospheric Pressure: Weight of air at sea level, about 14.7 pounds per square inch (101 kPa), decreasing at higher altitudes.

Backlash: A lack of mesh between two gears, resulting in a lag between when one gear moves and when it engages the other.

Base Circle: An imaginary circle drawn on the profile of a cam lobe that intersects the very bottom of the lobe. It is the lowest point on the cam lobe in relation to the valve train.

Bend: On a connecting rod, the condition of the two bores being out-of-parallel when viewed from the edge of the rod.

Boost Pressure: The amount of air pressure increase above atmospheric pressure provided by a turbocharger.

Carbon Monoxide: An odorless, colorless, tasteless poisonous gas. A pollutant produced by an internal combustion engine.

Chamfer: A beveled edge.

Compression Pressure: The total amount of air pressure developed by a piston moving to TDC with both valves closed.

Concentricity: Having the same center, such as two circles drawn around a common center point. A valve guide and valve seat must be concentric, or have centers at the same point.

Core Plug: A shallow, metal cup inserted into the engine block to seal holes left by manufacturing. Also called a freeze plug or expansion plug.

Crosshatch: A multi-directional surface finish left on a cylinder wall after honing. The crosshatch finish retains oil to aid in piston ring seating.

Data Link Connector: An electrical connector that allows a scan tool to communicate with the powertrain control module (PCM).

Eccentric: Off center or out of round. A shaft lobe which has a center different from that of the shaft.

Electrolyte: The chemical solution, generally sulphuric acid and water, in a battery that conducts electricity and reacts with the plate materials.

Endplay: Movement along, or parallel to, the centerline of a shaft. Also called end thrust or axial play.

Expander Broach: A tool used to seat a bushing and form the outside diameter of the bushing to the irregular surface of the bore.

Fillet: A curve of a specific radius machined into the edges of a crankshaft journal. The fillet provides additional strength between the journal and the crankshaft cheek.

Frequency: The number of periodic voltage oscillations, or waves, occuring in a given unit of time, usually expressed as cycles per second or Hertz.

Hall-effect Switch: A semiconductor that produces a voltage in the presence of a magnetic field. This voltage can be used to cycle a transistor on and off to produce a variable-frequency digital signal.

Helical Insert: A precision-formed coil of wire used to repair damaged threads, or to reduce the internal diameter of a bored hole.

Hydrocarbon: A chemical compound of hydrogen and carbon. A major pollutant from an internal combustion engine. Gasoline itself is a hydrocarbon compound.

Hydrometer: A device used to measure the specific gravity of a fluid.

Hypereutectic: A casting process that combines aluminum with small silicon particles. The silicon particles provide a durable surface finish.

Installed Height: The dimension from the valve spring seat to the bottom of the spring retainer; also called assembled height.

Integral: A part that is formed into a casting.

Interference Angle: The difference between the angle at which the valve face is ground and the angle at which the valve seat is ground.

Manifold Pressure: Vacuum, or low air pressure, in the intake manifold of a running engine, caused by the descending pistons creating empty space in the cylinders faster than the entering air can fill it.

Micropolishing: A machining technique that uses an abrasive belt to restore mild crankshaft journal damage while removing minimal amounts of metal.

Milling: The process of using a multiple-tooth cutting bit to remove metal from a workpiece.

Offset: On a connecting rod, the condition of the two bores being out-of-parallel when viewed from the side.

Oil Dilution: Oil thinned or diluted by unburned fuel which gets past the piston rings and into the crankcase. Oil dilution also can result from water condensation in the crankcase.

Oil Galleries: Pipes or drilled passages in the engine block that are used to carry engine oil from one area to another.

Open-loop: An operational mode in which the engine management computer adjusts the system to function according to predetermined instructions and does not respond to feedback signals from system input sensors.

Oscillating: A swinging, steady, up-and-down or back-and-forth motion.

Parting Line: The meeting points of two parts or machined pieces, such as the two halves of a split shell bearing.

Plateau Surface: A finish in which the highest points of a surface have been honed to flattened peaks.

Ported Vacuum: The low pressure area, or vacuum, located just above the intake air throttle plates of an engine.

Positive Displacement Pump: A pump that delivers the same volume of fluid with each revolution regardless of speed.

Reamer: A side-cutting tool used to finish a drilled hole to an exact size.

Residual Pressure: A constant pressure held in the fuel system when the pump is not operating.

Resistance: Opposition to electrical current.

Ridge Reamer: A hand-operated cutting tool used to remove the wear ridge at the top of a cylinder bore.

Ring Lands: The part of a piston between the ring grooves. The lands strengthen and support the ring grooves.

Rodding: A method of repairing a radiator by running a drill-bit through the internal passages of the core.

Runout: Side-to-side deviation in the movement of a rotating assembly.

Schrader Valve: A service valve that uses a spring-loaded pin and internal pressure to seal a port; depressing the pin will open the port.

Signal Generator: An electrical device that creates a voltage pulse that changes frequency or amplitude, or both, in relation to rotational speed.

Sludge: Black, moist deposits that form in the interior of the engine. A mixture of dust, oil, and water whipped together by the moving parts.

Specific Gravity: The weight of a volume of liquid divided by the weight of the same volume of water at a given temperature and pressure. Water has a specific gravity of 1.000.

Stretch: An extreme out-of-round condition on a bearing bore.

Supercharger: A crankshaft-driven compressor that delivers an air-fuel mixture to the engine cylinders at a pressure greater than atmospheric pressure.

Surface Grinding: The process of using a power-driven abrasive stone to remove metal from a casting to restore the surfaces.

Thermostatic Bulb: A device that automatically responds to changes in temperature to actuate a damper in the air intake passage.

Throating: Term for raising a valve seat in the cylinder head and narrowing the valve-to-seat contact patch by grinding an angle 15 degrees greater than the seating angle.

Tipping: Term for grinding the stem end of a valve to maintain correct stem height after grinding the valve face.

Topping: Term for lowering a valve seat in the cylinder head and narrowing the valve-to-seat contact patch by grinding an angle 15 degrees less than the seating angle.

Turbocharger: A compressor device that uses exhaust gases to turn a turbine that forces the air-fuel mixture into the cylinders.

Twist: On a connecting rod, the condition of the two bores being out-of-parallel when viewed from the top.

Valve Duration: The number of crankshaft degrees that a valve remains open.

Valve Lash: The gap between the rocker arm or the cam follower and the tip of the valve stem when the valve is closed. Also called valve clearance.

Valve Margin: The distance on a valve from the top of the machined face to the edge of the valve.

Valve Timing: A method of coordinating camshaft rotation and crankshaft rotation so that the valves open and close at the right times during each of the piston strokes.

Valve Train: The assembly of parts that transmits force from the cam lobe to the valve.

Variable-pitch Spring: A spring that changes its rate of pressure increase as it is compressed. This is achieved by unequal spacing of the spring coils.

Varnish: A hard, undesirable deposit formed by oxidation of fuel and motor oil.

Voltage Drop: The measurement of the loss of voltage caused by the resistance of a conductor or a circuit device.

Wastegate: A diaphragm-actuated bypass valve used to limit turbocharger boost pressure by limiting the speed of the exhaust turbine.

Water Jacket: The area in the block and head around the cylinders, valves, and spark plugs that is left hollow, so the coolant can circulate.

Wrist Pin: The cylindrical, or tubular, metal pin that attaches the piston to the connecting rod (also called the piston pin).

Chapter 1:

1. d, 2. b, 3. c, 4. b, 5. c, 6. b, 7. d, 8. a, 9. b, 10. c

Chapter 2:

1. b, 2. d, 3. c, 4. d, 5. a, 6. d, 7. c, 8. a, 9. c, 10. c, 11. b, 12. d, 13. c, 14. b, 15. a

Chapter 3:

1. c, 2. b, 3. c, 4. d, 5. b, 6. a, 7. c, 8. d, 9. a, 10. b, 11. d, 12. c, 13. d, 14. b, 15. b

Chapter 4:

1. c, 2. a, 3. c, 4. d, 5. b, 6. c, 7. a, 8. b, 9. d, 10. b

Chapter 5:

1. a, 2. b, 3. d, 4. b, 5. c, 6. c, 7. a, 8. a, 9. b, 10. b, 11. c, 12. b, 13. d, 14. b, 15. d, 16. b, 17. c, 18. c, 19. a, 20. b

Sample Test:

1. b, 2. c, 3. c, 4. a, 5. c, 6. c, 7. d, 8. b, 9. b, 10. d, 11. b, 12. b, 13. d, 14. d, 15. d, 16. a, 17. b, 18. d, 19. c, 20. a, 21. c, 22. b, 23. d, 24. b, 25. a, 26. b, 27. c, 28. c, 29. d, 30. c, 31. b, 32. d, 33. b, 34. d, 35. c, 36. b, 37. b, 38. c, 39. b, 40. c